// Also by Russell Gold

THE BOOM
How Fracking Ignited the American Energy
Revolution and Changed the World

SUPERPOWER

One Man's Quest to
Transform American Energy

RUSSELL GOLD

Simon & Schuster

New York London Toronto Sydney New Delhi

Simon & Schuster
1230 Avenue of the Americas
New York, NY 10020

First Simon & Schuster hardcover edition June 2019

SIMON & SCHUSTER and colophon are registered trademarks
of Simon & Schuster, Inc.

For information about special discounts for bulk purchases,
please contact Simon & Schuster Special Sales at 1-866-506-1949
or business@simonandschuster.com.

The Simon & Schuster Speakers Bureau can bring authors to your
live event. For more information or to book an event, contact the
Simon & Schuster Speakers Bureau at 1-866-248-3049 or visit our
website at www.simonspeakers.com.

Interior design by Lewelin Polanco

Manufactured in the United States of America

10 9 8 7 6 5 4 3 2 1

Library of Congress Cataloging-in-Publication Data
Names: Gold, Russell, author.
Title: Superpower : one man's quest to transform American energy /
    Russell Gold.
Description: First Simon & Schuster hardcover edition. | New York :
    Simon & Schuster, [2019] | Includes bibliographical references.
Identifiers: LCCN 2019015504| ISBN 9781501163586 (hardcover) | ISBN
    9781501163593 (pbk.) | ISBN 9781501163609 (Ebook)
Subjects: LCSH: Skelly, Michael, 1961- | Energy industries—United
    States—Biography. | Businesspeople—United States—Biography. |
    Energy policy–United States.
Classification: LCC HD9502.U52 S634 2019 | DDC 333.79/4092 [B]—dc23
LC record available at https://lccn.loc.gov/2019015504

ISBN 978-1-5011-6358-6
ISBN 978-1-5011-6360-9 (ebook)

For L.G. and R.G.,
who have renewed me

"While everybody talked about the weather, nobody seemed to do anything about it."

—CHARLES DUDLEY WARNER,
Hartford Courant, 1897

Contents

SUPERPOWER

// **1** //

No Man's Land

When Michael Skelly first visited the Oklahoma panhandle in the summer of 2009, he had a good idea of what he would find. He expected lots of sunbaked and windswept open space, and he wasn't disappointed.

It was a hot day as he drove west from Guymon, the largest town in the panhandle, into a landscape that was pancake flat and dry. There were only a handful of trees. For mile after mile, the two-lane Highway 3 ran straight as a matchstick. Every mile, a dirt road intersected the highway and headed off toward the horizon at a 90-degree angle. The land was a giant grid containing squares of corn, milo, and grassland. There were few houses, one every mile or so. Half had been abandoned decades ago by homesteaders who gave up their fight with the elements.

Skelly didn't want to fight the elements. He wanted to harness them.

Gazing at the landscape, Skelly grew excited. He knew no place is perfect for energy development. But the panhandle was quite good. There was lots of room to build, thousands of square miles. Local landowners were supportive. No endangered animals lived there. You could build renewable energy here, he thought, on a scale that could change the country and maybe even the planet.

The wind rarely stopped blowing. A bestselling midcentury travelogue joked that homes here had a "crowbar hole . . . designed to check on the weather. You shove a crowbar through the hole: if it bends, the wind velocity outside is normal; if the bar breaks off, 'it is better to stay in the house.'" A few years earlier, Skelly's former employer, a wind farm developer, had erected a couple of needle-thin meteorological towers with instruments to measure wind speed. They were the tallest structures for miles around. The results were striking. The winds were strong and surprisingly steady. The data collected was good enough to go to a bank and get financing for a wind farm, but Skelly had never done anything about it. The region was too remote. Sure you could generate a lot of power, but then what? Where would it go?

The sun was also relentless. On the giant interconnected power grid that runs from Nova Scotia, along Canada's Atlantic coast, down to Miami and across to Montana, the sunlight is strongest in the Oklahoma panhandle and an adjacent area in New Mexico. If you put identical solar panels in the westernmost counties of Oklahoma and in Miami, you would generate one-third more electricity in the panhandle—and nearly twice as much as in Cleveland, Pittsburgh, or Albany. In the 1970s, the *Guymon Daily Herald* ran a box on the top left of the paper every day that proclaimed the city was the "Sunshine Capital of the World."

At the time, Skelly was in his late forties. He was the human equivalent of a meteorological tower: slightly more than six feet tall and fit from years of biking to and from his office in Houston. Skelly had bushy hair and wore browline glasses. He was a departure from an earlier generation of wind farm builders and solar panel enthusiasts who were more interested in creating sustainable energy than sustainable businesses. Skelly wanted to make a profit, because profits would attract new investors and money into renewable energy.

Traveling around the panhandle, Skelly allowed himself to think big. He envisioned an energy development unlike anything ever built before. It would be enormous, and would generate power at a cost lower than anything that had come before it. And because the sun and the wind were so steady, the power would be available, on average, nearly twenty hours a day.

The panhandle had a lot to recommend it as a place to build renewables. But there was a big drawback. No one nearby needed the power. Skelly planned to solve this problem by building a long extension cord: one end would be plugged into the panhandle; the other end would reach east until it crossed the Mississippi River. It would be the first of a set of transmission lines he wanted to build that would carry current hundreds of miles from one state to the next, stitching together the country and delivering low-cost, carbon-free electricity. He would upgrade the existing power grid, a century-old engineering marvel that tied together power plants burning a lot of fossil fuels. "It's an environmentally reckless business model in need of reimagining," he thought. Skelly dreamed of a new grid that could power modern society without contributing to the darkening cloak of carbon dioxide around the Earth.

Of course, such a grid would cost billions and billions of dollars. Skelly wanted to show it could be done, and done profitably. It was ambitious but necessary, he figured. "If we don't do it," he thought to himself, "who is going to do it?"

A year before Skelly's visit, Carroll Beaman had also concluded the Oklahoma panhandle was a good place to erect wind turbines.

He was born on an infamous day in the region's history. "I came in with a storm," he said. Beginning in early 1932, enormous dust storms blew across the Oklahoma panhandle and neighboring states. The wind scooped up millions of tons of topsoil that homesteaders had plowed up in their misguided effort to turn well-adapted grassland into fields of wheat. After a few years of this new agricultural practice, there were no more roots to hold the soil in place. A strong wind would create giant moving clouds called "dusters." The region became known as the Dust Bowl. It was one of the worst man-made environmental disasters ever, and it led to mass migration.

The first giant dust storm was on January 21, 1932. Beaman was born that day. His parents were homesteaders, he said, and "just as poor as anybody else." His mother was born in a dugout her parents had carved into the dirt. His grandfather raised crops and owned a water-drilling rig. The contraption would bore fifty feet, sometimes deeper, into the red dirt until it found water. On top of the borehole, the farmer would install a windmill, an Aermotor or a Dempster, to pump up water for cattle or crops.

"Dusters" are now part of history. New farming practices and countywide soil conservation districts ended the era of stripping away prairie grasses. Human activity had destroyed the ecosystem, but in time humans restored a semblance of balance. The weather had once brought misery to the panhandle. Now Beaman wanted the wind to bring investment and steady checks.

He split his time between a home in Amarillo, Texas, a two-hour drive to the south, and a second house in Guymon. He came to the panhandle in Oklahoma as much as possible to tend his garden of cucumbers, asparagus, and tomatoes next to his family's homestead nearby. One day in 2008, driving north from Amarillo, he noticed large cranes assembling gently tapered steel tubes. On the return trip, three long aerodynamic blades had been attached to a large box on the top of each tube. It was the first time he had seen a modern wind turbine. He thought the machines looked solid and impressive. Here on the southern edge of the Great Plains, grain elevators seemed as tall as skyscrapers. But these wind turbines were twenty and thirty stories tall; they dwarfed the grain elevators. They were nothing like those twenty-foot windmills dotting the region that were rusting in place. The new ones were elegant machines—two rows of eight turbines each eventually straddling the rural road. His grandfather's windmills once pumped water up from the aquifer; these new turbines were for electricity—wind turned the blades, spinning magnets that generated currents to be fed into the grid.

"We have some of the best winds in the world," he thought on a drive home to Amarillo. "There is no reason this area wouldn't be a prime candidate for development."

While in his car, Beaman also ran some numbers in his head. "I saw these groups of eight turbines. They are on a quarter section, half section," he said. Sections are 640 acres of land, one mile wide and one mile long. "I saw those and thought, hell, if eight turbines are economic, then what I can do is put eight turbines on every half section I got, tie it into an electric line, and that is all there is to it."

Years later, he chuckled. "I found out there was a little more to it."

―――――――――

After seeing the new turbines, Beaman drove around the counties in Oklahoma's panhandle, talking to his neighbors about building a wind farm. He knew most of the people who lived in the panhandle and had known them for decades. He urged them to sign energy leases. They would pool their land and use their size to get the best bargain.

The western three counties that form the state's thin western appendage are called No Man's Land. Between 1850 and 1890, no state claimed the region. It was an empty spot on the map and remains sparsely populated. In the 1990s, a giant pork conglomerate moved in and built a slaughterhouse on the outskirts of Guymon. About twenty thousand pigs arrive in trucks every day from barns scattered across the panhandle and neighboring states. By one estimate, there are sixty-two hogs for every person in the counties that make up the panhandle.

After attending college in Colorado, Beaman ended up working for Exxon overseas in the 1950s and 1960s. He returned to Amarillo to run a small oil and gas company, socializing with the city's business and civic leaders including fellow oilman T. Boone Pickens. One of the elementary rules of oil exploration is that when you make a discovery, lease up as much surrounding land as possible. Beaman took the same approach to wind.

Driving around the flatlands, he discovered that he didn't have to make a hard pitch to get his neighbors to sign up. One rancher "signed the lease, never read it," he said. "He trusts me." People understood that the wind turbines wouldn't interfere with farming. The turbine's concrete bases could be put in the corner of each grid square, just like the hog barns. Beaman promised his neighbors payments down the road when deals were made.

Out in No Man's Land, where the thin Oklahoma panhandle nestles up against Kansas, Colorado, Texas, and New Mexico, Beaman began to assemble commitments from his neighbors. Beaman, who often wore a cream-colored cowboy hat and oversized belt buckle won in a cutting horse riding competition, called the company CimTexCo, for Cimarron and Texas Counties. Beaman soon learned that there was a reason those first wind turbines he had seen on his drive from Amarillo were located where they were. They were close to power lines. To make CimTexCo viable, he needed a power line. He put in a proposal with the local grid operator to extend a transmission line west into the panhandle. The Southwest Power Pool assigned it to its priority project list. It was number 13, so far down the list he doubted he would be alive to see it completed.

His frustration was building when a mutual acquaintance introduced him to Michael Skelly, and he learned about the interstate extension cord idea. They were working on complementary plans.

Beaman realized his best chance to bring investment to the panhandle was if Skelly's transmission line could be built. Beaman increased his outreach and soon had 100,000 acres committed to CimTexCo. This would eventually grow to more than 300,000 acres—about 500 square miles. It was enough land for strings of turbines to extend for sixty miles, east to west. It was enough to build the largest renewable energy project on the continent.

One day he was at his lawyers' offices, working on the paperwork for CimTexCo wind leases. He marveled at the size of the project, and so did an out-of-town lawyer. "None of us have ever dealt with anything of this big a scale or magnitude. It is uncharted territory for us all," Beaman said.

But without Skelly's extension cord, none of it made a lick of financial or engineering sense. Beaman needed the line. "I am banking on it," he said.

If Skelly could build a transmission line from Oklahoma to Memphis, Tennessee, the panhandle would sprout two thousand wind turbines and tens of thousands of solar panels. The electricity generated would flow into a substation near Guymon and then on to a high-voltage direct current express line headed east. It would drop off some power in Arkansas, and take the rest over the Mississippi River into Memphis. From there, the flow of electrons that make up an electrical current would be on a grid operated by the Tennessee Valley Authority. In the 1970s, the federal agency planned to build more than a dozen nuclear power plants in the valley. The TVA built the wires but then scrapped plans for the nukes. If Skelly could get the power across the Mississippi into Memphis, he could move the power north, east, and south on this oversized network. Atlanta, Charlotte, even Washington, D.C., and from there Philadelphia and New York would be within reach.

It was such a bold idea that some people in the Oklahoma panhandle had trouble believing it. Jay Lobit's initial reaction was that it sounded preposterous. He had returned to the area in 2005 to build wind farms because of his recollection of working as a paperboy for the afternoon newspaper and getting dragged across Main Street in Guymon when a gust of wind turned his canvas delivery bag into a sail. A private line all the way to Memphis? "Yeah, right," he thought. But he was curious. "All right," Lobit told a friend who wanted to introduce him to Skelly. "I'll go meet this nut." He wanted to see what kind of masochist would attempt such a project.

They met at a conference in Oklahoma City in 2009 and chatted for an hour. Skelly was about twenty years younger and a head taller than Lobit. Skelly spoke passionately about his idea. Lobit felt himself getting drawn in. Lobit still believed an Oklahoma-to-Tennessee power line was insane, but if anyone could pull it off, Lobit was prepared to bet on this goofy but earnest guy. Skelly had a "grand vision," Lobit said, that would boost renewable energy and the fortunes of renewable energy developers like him. "He talked about his plan and what it was going to be. I was happy to have any kind of a plan," Lobit remembered.

Skelly wanted to permit and build a 720-mile electricity expressway through the middle of the country, bypassing a snarl of local and often congested power lines. His proposed power line had the capacity to carry 4,000 megawatts of wind and solar power. This was enough to carry electricity from Beaman's CimTexCo, the wind company Lobit had cofounded, and others. Adding them all up, the panhandle wind and solar farms would be one of the largest power-generating projects in the United States. It would trail only the gigantic Grand Coulee Dam, which sends the gravity-propelled Columbia River through twenty-five generators and produces up to 6,809 megawatts, and the Palo Verde nuclear power plant west of Phoenix, which has a peak output of 4,219 megawatts. Here's another way of understanding the scale of Skelly's vision: One of the United States' leading rooftop solar panel companies, SolarCity, was founded in 2006 and became part of Tesla and Elon Musk's empire a decade later. As of early 2018, it had installed 3,310 megawatts. Skelly's Oklahoma panhandle renewable energy complex would be bigger and produce power at a much lower cost.

Capacity can be misleading. A typical car speedometer might say it is capable of going 140 miles per hour, but most cars never go near that maximum speed. A car's top speed is a bit like installed capacity for a wind or solar farm. A better way to think about power plants is in how many megawatt hours are produced, comparable to a car's average speed. By that measure, the three largest power plants in the United States are all nuclear facilities: Palo Verde in Arizona, Browns Ferry on the Tennessee River in northern Alabama, and Oconee in the northwest corner of South Carolina.

Right behind them would be the proposed cluster of wind and solar farms in the Oklahoma panhandle, churning out electricity that travels on the new grid to faraway cities.

Skelly's plan was ambitious. But the reward at the end of the line was significant: it was a huge step forward in a transition to a cleaner energy future. It would also be a big payday. Changing the electricity system would be enormously expensive. But if Skelly could set the precedent with a moneymaking Oklahoma-to-Tennessee line, it would be worth the headache.

Skelly left No Man's Land in the summer of 2009 with a single-minded focus. The counties were big and windy and sunny, and there were developers champing at the bit. A self-described "infrastructure nerd," Skelly wanted to build a massive energy project that was both inspirational and profitable. This was how the energy transition would happen, he figured. One step at a time and one transmission line at a time.

Over the next few years, Skelly didn't get involved in Twitter feuds or policy battles. He wasn't interested in wasting energy fighting over energy. Why bother, he would ask with a shrug. He much preferred to use his time developing energy projects.

For Skelly, the energy transition wasn't an idea to be debated. It was something that needed to be financed, planned, and built. The longer it took to rewire the country, the more heat-trapping gases would be released into the atmosphere. That meant more climatic changes that could make the Oklahoma Dust Bowl pale in comparison. The world was getting more dangerous. Storms were stronger and droughts more persistent. All this was happening and the global mean temperature had only risen by less than one degree Celsius. What would happen if temperatures rose by two degrees? Or three?

Skelly had spent a lot of time in the year before his visit to Oklahoma reading up on climate change and thinking about it. He found it depressing and distracting. He much preferred doing something about it. The transmission line would be extraordinarily difficult, but he thought it was a good way to spend a few years. It was challenging, but worthwhile.

"This is the business of overcoming obstacles," Skelly said. "It is not like you just wake up in the morning and all of a sudden, you know, the heavens part and you get to go build the project." Skelly set out to clear a path through the obstacles.

2

E Pluribus Unum

M ichael Skelly, the son of an electric engineer, has spent years building wind farms and plotting power lines. Yet he once admitted: "I don't really understand all the electrical stuff."

Neither do most people. It's a blessing of the modern age that we don't need to. When we flip a switch, the light goes on. It stays on as long as we want it, steady and constant without flickering or surging.

Skelly is intelligent and ambitious, idealistic and profane. He is a city dweller, but his near-daily forays into the Houston's parks and bayous on bicycle allow him to see the city with a naturalist's curiosity. Underneath a veneer of wit and self-deprecating humor, he takes himself seriously. When he says he doesn't understand the electrical stuff, it is wise not to believe him. Perhaps he doesn't know the details of the power grid with the precision of an electric engineer. But he has a good sense of the political, social, environmental, and economic forces at work.

He lives in a part of Houston called the East End. It is mostly smaller, older homes that date from when the neighborhood sprang up around a Ford Motor Company factory. Skelly's home, which he shares with his wife, Anne Whitlock, is the former Houston Firehouse No. 2. There are two brass poles connecting the second floor to the first. See-through covers over the holes prevent tipsy visitors from sliding down and twisting an ankle. Exceptions are made. At a political fundraiser, Beto O'Rourke entered by sliding down a pole into a crowd of donors.

When the firehouse was built, in 1910, a single power plant served all of Houston. A two-story long brick building, the Gable Street plant was located along the Buffalo Bayou on the outskirts of downtown. In 1898, the facility was the site of a major accident. "Fatal Boiler Explosion," screamed a headline in the *Houston Daily Post*. Two men died and three were badly injured. With the plant out of commission, the paper noted, "the city was plunged into total darkness." After the accident, a new power company emerged and a new power plant rose from the debris. As the city grew, Houston Lighting & Power Co. added steam engines and turbines to the plant. In 1910, it could generate 7.2 megawatts. A home or business paid about 10 cents for every kilowatt hour. If they used a lot, the bulk rate went down to 7 cents.

As the crow flies, power from the Gable Street plant traveled a little more than a mile to the firehouse. Today, power flows across the state in a giant power network, a sinewy web of electricity. It is easier to find out where the apples in the grocery store were grown than where the power in the wall socket originated. Even the people who administer the state's power grid have trouble answering a question of provenance.

Electricity is electricity, whether it comes from the large nuclear plant an hour's drive down the coast from Houston, the wind turbines along ridges in West Texas, gas plants to the east of the city, or the giant coal plant about an hour's drive northwest of Houston that releases a thick dull gray smoke that is visible for miles against the blue sky. Electricity flows from many places, moving along copper or aluminum wires to power modern life: night-lights in children's rooms, massive steelmaking electric arc furnaces, and data centers that provide instant access to YouTube videos and Netflix movies.

Skelly's home in Houston sits near the southeastern edge of a large power grid that covers most of the state and goes by an unwieldy name: the Electric Reliability Council of Texas. Everyone calls it ERCOT. Despite its size—six hundred miles east to west and even larger north to south—it is considered a minor grid, especially compared to its two giant neighbors, the Eastern Interconnection and the Western Interconnection.

The Western Interconnection covers El Paso to Vancouver, sweeping up everything west of the Rockies. The Eastern Interconnection is even larger, connecting Toronto and Miami in its wiry web. These grids serve a basic, important function. They provide a reliable, consistent source of power, so when you flip on the light, there is current available at a consistent voltage level. The grid managers are like Goldilocks looking for the porridge that is neither too hot nor too cold. Both of the two big grids are so large they contain about three dozen "balancing authorities" responsible for keeping the grid in constant equilibrium.

From a control room located thirty miles northeast of Austin in an unmarked bunkerlike building behind a high black metal fence, ERCOT balances its entire grid. Across from its front gate is a small airport. You could park your car on the shoulder and walk onto the runway. Getting into the bunker requires negotiating three distinct layers of security. To enter the innermost sanctum, the darkened control room, employees have to place their thumbs on a fingerprint scanner.

On the day I visited the bunker, there were eight men wearing jeans and untucked shirts, sitting at workstations. On the wall, a thirty-foot-wide screen showed the Texas power grid, color-coded to indicate whether any transmission lines were down and a hundred other details, including the weather and local road congestion. There could be a tornado outside and the bunker would be quiet and unaffected.

There are a thousand tasks these men must attend to during an eight-hour shift, but only one job: to keep the grid running, around the clock, regardless of whether the temperature is in the triple digits or there's an ice storm. Dan Woodfin, the director of system operations, said he hired a lot of former military and linemen who worked on the wires. They were calm under pressure and accustomed to following rules precisely.

"We have to exactly balance generation and load all the time," said Woodfin. He dressed in a white shirt and black slacks, with combed-back hair and no-nonsense glasses. He wore a single ring on each hand: one from marriage, the other from college. Generation means the power supply. Load is the term used in the electricity business to indicate power consumption. When you turn on a lamp, you are increasing the load. When a coal plant fires up, or a solar panel catches the sun, that's an increase in generation.

All of the power grids in North America aim to have the electricity in their system oscillating at exactly 60 hertz. That means that sixty times a second an electromagnetic wave carried by wires hits a peak. But these waves are fickle. If there is more generation than load, the frequency speeds up. A large display in the ERCOT control room shows the frequency at all times. If the load increases quicker than generation, the electromagnetic wave elongates and the frequency slows down. The frequency is a bit like a car tire out of balance. If you drive at a certain speed, everything will feel fine. But go too fast or too slow, and the tire wobbles.

Struggling to understand, I suggested an analogy to Woodfin. The grid is like a water reservoir that you want to keep at an exact level. There are numerous places where water can be added, and an even larger number of spillways and buckets taking water out. The balancing authority monitors the water level, taking steps to keep the inflows and outflows constant.

"Water moving in and out of a lake is a slow motion thing relative to this, which is instantaneous," Woodfin said. In other words, imagine a reservoir that could drop—or rise—by several feet in an instant. And imagine if that sudden movement interfered with the operation of ventilators for patients. My analogy lacked urgency. I hadn't grasped the life-and-death aspect of his work.

There's a large map of Texas on the wall of Woodfin's office. Something about it didn't make any sense and appeared deeply confused. The cities were in all the right places, but the road system was off. There were thick red lines headed into Houston that might have been interstates but at odd angles. There was a large loop road that circled the city, and dozens of smaller feeder roads coming into and out of the thick red lines. It was a map of the Texas power grid. I was seeing the state in an entirely different way. It's like the beginnings of a riddle: What is everywhere, but never seen?

The map reminded me of a conversation I had a few months earlier. I was visiting an environmentalist named Bob Allen who lives in the middle of Arkansas in a home he built himself. It is a beautiful and rural region, just south of the Ozark National Forest. He often sees bobcats and otters from his wooden deck. "It is a very rugged, isolated area," he said. "We are a long way from anywhere." His house was between the Oklahoma panhandle and Memphis, and Skelly's transmission line would likely run within a couple miles of his home. Allen was in favor of the line because he wanted more renewable energy.

Power lines, he said, "are unsightly, there is no doubt about it." But that is only when you go looking for them. Mostly, he said, they are unseen. One day, on the twenty-mile drive from his home to the nearby city of Russellville, he ran a test. "I counted the number of power lines I drove under. Over one hundred from here to Russellville. They are so common, they are essentially invisible to us these days."

I had driven out from Russellville that morning and had trouble believing him. There was no way I had driven under one hundred power lines. But as I pulled out of his driveway, past his solar panels on the right and a meadow holding a flock of lambs on my left, I kept a tally in my head. I had gone under four power lines in the half a mile on a dirt road back to the main state highway. By the time I had driven four miles and passed a sign for the Booger Hollow Tabernacle, I had lost count. On my way to visit him, I hadn't noticed any of them. They were such a part of the modern landscape that even in rural Arkansas their existence hadn't registered with me.

To move power around the country, there are about 160,000 miles of high-voltage power lines. These are the large lines that hang off fifteen-story lattice towers. That is more than three times the size of the U.S. interstate system. Unlike highways, which tend to connect cities with each other, the U.S. high-voltage power lines connect the machines that generate power with cities and industries that use it. Millions of miles of smaller low-voltage lines connect neighborhoods and houses.

In 2003, the National Academy of Engineering presented its choice for the greatest achievements of the twentieth century. It listed cars and planes, computers and the internet. Above them all was electrification: the power grid. We accept dropped calls and mobile phone apps that crash. This level of service will not do for grid operators. They know the exact date of the last rolling brownout. Investigations into widespread blackouts often end careers.

This fear of failure is why grid operators worried when renewable energy began to grow. The power grid had run on dispatchable power: Power plants offer their services. They send in computerized bids that say, essentially, we can provide 300 megawatts of power from 3 p.m. to 4 p.m. tomorrow and we're available if the price is at or above $30 per megawatt hour. If the grid operator accepts the bid, the power plant gets a dispatch ticket, and it is expected to be spinning its turbines and producing 300 megawatts when the clock strikes 3 p.m.

But renewable energy is different. The eight men in the ERCOT control room can't tell a wind farm to provide power if the wind isn't blowing. When Woodfin arrived at ERCOT in 2003, there were 1.2 gigawatts worth of wind power in ERCOT. Back then, if *all* of the turbines in ERCOT were to run at maximum output, they would have equaled one of the units at the South Texas Project, a nuclear plant on the flats near the Gulf of Mexico. "I remember folks saying, at that point in time, that we would have to do some things radically different if we got about 15 gigawatts," Woodfin said. But Texas reached 15 gigawatts, or 15,000 megawatts, in 2015, and the sky didn't fall. Wind kept growing. ERCOT got very good at day-ahead wind forecasting. Wind didn't become dispatchable, but it became predictable. And while the system had been built on dispatchable power, it was also built for a large amount of chaos. Sometimes, nuclear power plants must shut down quickly if there's a safety problem. Giant industrial machines can fire up unexpectedly. The grid operators must be prepared.

On March 23, 2017, wind farms provided 50 percent of all the electricity in ERCOT for the first time. It came at 4 a.m. on a Thursday morning. There were no panicked calls to Dan Woodfin's cell phone. It was just another day on the Texas grid. It was the first time, but not the last that renewables crossed this threshold.

The ability to handle this much wind surprised grid engineers. In 2017, Nick Brown, president of the Southwest Power Pool, a grid operator similar to ERCOT but for several Great Plains states, testified before Congress that it was handling 17,000 megawatts of wind without any issues. "I will tell you as an engineer, with training in operations and planning, if you had asked me 10 years ago if we would have been able to reliably accommodate even half of that, I would have said no. Period. End of discussion," he said.

In 1898, the failure of a single power plant in Houston brought darkness to the whole city. Today, the grid can compensate and accommodate fluctuations. Working in concert, thousands of generators and lines keep the grid at 60 hertz. Getting to this point took decades of investment and created a system of powerful industry insiders who controlled large portions of the grid. An outsider such as Skelly wouldn't be easily welcomed into the club.

———————

The modern power industry began in a single factory that belonged to Joseph E. Hinds. Born in Brooklyn in 1848, he began working at eleven years old and found his way into printmaking at fifteen. The companies he worked for made colored labels to affix to manufactured goods. By the time he was thirty, he had his own business.

In October 1880, he was working in the offices of his new company, Hinds, Ketcham & Co., when a visitor came with questions: How many gaslights did the firm use? How much did he pay for the gas? Hinds answered and then asked the stranger a few questions of his own. Why do you want to know? Who are you working for?

The visitor explained that he was surveying lower Manhattan for Thomas Edison, who was working on a new form of lighting. Hinds had heard of Edison. A few months earlier, an illustrated newspaper had called him a "Wizard of Electricity" whose invention provided light with "no deleterious gases, no smoke, no offensive odors." Hinds traveled to Edison's New Jersey laboratory to make the inventor an offer. He was building a new factory for his printing presses and label makers. Did Edison want to use it to test this new lighting apparatus?

Edison agreed and sent over a small generator and wired Hinds's factory in a jury-rigged fashion. Lacking the proper insulation, workers tacked naked copper wires onto the walls. By the end of December or perhaps early in 1881, the new system was ready for a test run. When Hinds leaned over to throw the switch, his coat caught in a belt and he was thrown to the floor. It was an inauspicious beginning. Hinds picked himself up—and threw the switch. The lights went on. Electricity lit its first business.

Hinds printed up a circular announcing that a "special dynamo electric machine" powered his new business on Water Street. Other label makers who worked at night with candles or gaslight struggled, he said, but the new incandescent bulbs were a "true substitute for daylight, showing all colors in their natural hues." Curious crowds streamed through the factory.

A few months later, Edison supervised another installation. This time, the recipient was his financial supporter and one of the wealthiest and most powerful men in the world: J. P. Morgan. The tycoon was renovating his Madison Avenue three-story brownstone and decided he would show off Edison's invention by adding electric lights. He installed a steam engine, boiler, and two generators in a room underneath the house's horse stable. When the renovation was completed, in June 1881, every room had lights activated by turning a knob upon entry. An engineer showed up every day at three o'clock to feed coal into the boiler to generate steam so that the lights could go on at four o'clock. The engineer turned off the generator nightly at eleven. Inside the house, the lights were elegant and modern. Outside, the noise of the generator and smoke from the coal aggravated neighbors.

Hinds's business and Morgan's mansion were islands of electricity. The generators fed a thin and often fickle strand of power to a few rooms. Edison understood these islands were neither economic nor particularly reliable. His aim was to build centralized power plants to serve as many customers as possible.

Edison was plotting to build the world's first electrical network. By the end of 1881, Edison workers had unspooled fifteen miles of wires under the narrow streets of lower Manhattan. The next year, Edison's workers installed several coal-fired generators inside a building on Pearl Street, near the entrance to the nearly complete Brooklyn Bridge. The effort was "akin to venturing on an uncharted sea," Edison told a reporter years later. When the facility was switched on in September 1882, it provided power to eighty-five customers and lit four hundred lamps in and around the city's financial district. By the beginning of October, wiring reached 1,600 lamps. A year later, the Pearl Street station could light 11,555 lamps. This rapid growth might have continued were it not for an 1890 fire that gutted the world's first power plant.

By the end of the decade, there were a thousand centralized power plants across the country, feeding power throughout neighborhoods via growing networks of wires.

Only a few years after New Yorkers marveled at Edison's lights, they began to grouse about the profusion of poles and overhead wires carrying the electricity. In 1889, the New York City mayor, a Tammany Hall Democrat named Hugh Grant, urged companies to move their wires from rooftops to underground conduits called subways. When the companies didn't act quickly enough, Grant armed crews of city employees with axes and sent them to chop down poles. They were "ready and anxious to start on a crusade against the copper and steel wire octopus that has so steadily encircled the streets in its tentacles," a newspaper reporter wrote. Crowds cheered as poles fell on Broadway.

Excited by success and possibility, a growing host of electric engineers set out to make power systems bigger and better. One challenge was to build a line that could carry power over long distances. In time for the International Electrical Exhibition being held in Frankfurt in 1891, a Berlin electrical manufacturer built a 108-mile line. A cement plant in Lauffen, Germany, would send its surplus power from a dam on the Neckar River. The faraway river's power flowed into Frankfurt and powered an artificial waterfall and a sign with a thousand bulbs that spelled out AEG, advertising the name of the transmission line's manufacturer.

Experts predicted the line would be a folly that would lose 50 percent of the power along the way due to resistance encountered along the metal conductor. But when calculations were run, the receiving terminal in Frankfurt received 74.5 percent of the power transmitted from the cement works. The success of the German line helped convince financiers to back a twenty-mile line connecting Niagara Falls to Buffalo. By 1895, long-distance, high-voltage power transmission was a reality. The lines grew longer. In 1901, a power plant that tapped the energy of the Yuba River, as it flowed from the highlands of the Sierra Nevada, supplied electricity to nearby gold mining operations, Sacramento, and the San Francisco Bay Area, via a 140-mile transmission line.

The long-distance line meant that power could be generated where it was convenient—where land was less expensive, where water was readily available, where plants could be built to a much larger scale. A 1902 survey of industry counted 3,620 central power stations and more than fifty thousand isolated power plants that served a single factory or a large home. Due to long-distance transmission, that lopsided equation would soon reverse.

Samuel Insull, a British immigrant who came to the United States to serve as Edison's secretary and later took over as head of Chicago Edison, recognized this trend. He helped speed the demise of the isolated power plant. In 1913, he gave a speech that described an experiment he had run several years earlier in Lake County, just north of Chicago. His company had purchased all the small power plants serving the area, including one run by a "gentleman farmer" who supplied power to some of his neighbors. Insull shut all of these plants down and replaced them with a single central plant along Lake Michigan. The cost of the new plant and miles of wires was substantial. But it allowed him to balance complementary demands for power. The new plant was more efficient and eliminated wasteful duplication. Despite the high construction costs, Insull lowered the price for electricity from 7 cents per kilowatt hour to 2.9 cents.

The conclusion was obvious to Insull and, he said, should also be to "any man of ordinary intelligence." All these small, isolated plants had to go and should be replaced by centralized power plants serving cities, or counties, or even larger regions of a state. But this wasn't Insull's final thought on the matter. If building regional power grids was logical and sensible, businesses that undertook the effort needed to be protected from fierce competition by being given state-sanctioned monopolies. Anything else, he sniffed, was "simply a waste of money."

———————

The savings and improved reliability meant it was remunerative to wire up a county. But why stop there? Why not connect an entire state? Why not connect the entire eastern seaboard? This was the logic of W. S. Murray, a skinny engineer with a Vandyke beard who wore a suit with a pocket square even when visiting a construction site. He was a towering figure. In a rare surviving photograph of him, the six-foot, seven-inch Murray towers over a fully grown man in a fedora. The tip of the hat doesn't quite reach the knot of Murray's necktie.

Murray had a fertile, adventurous mind. While not quite in the same league of electrical giants of the day—Edison, Tesla, Westinghouse, and Siemens—he was not far behind. The first time he demonstrated his ingenuity came after a spectacular train crash. On January 8, 1902, a passenger train was heading southbound in New York City under Park Avenue toward Grand Central Terminal in midtown Manhattan. Coal smoke from trains in the tunnel caused a conductor on the passenger train to miss a red signal and plow into another train, killing seventeen people. In response, before an inquest had even begun, politicians announced the trains needed to be electrified. "How it can be done is yet to be determined," said New York City mayor Seth Low.

State lawmakers soon banned trains from burning coal inside the city limits. Realistically, the only alternative was to use electricity. The law gave railroads until 1908 to figure out how to comply.

One of the companies that started working on this problem was the New York, New Haven & Hartford Railroad. Despite its provincial-sounding name, it was an enormous holding company owned by J. P. Morgan and William Rockefeller. It controlled the railroad between Boston and New York, as well as ten thousand other miles of tracks and coalfields in Pennsylvania and New York. In 1905, the railroad hired

Murray to be its chief electrical engineer and figure out how to run the trains effectively on electricity. Some city trolleys ran on electricity, but not any trunk line railroads that traveled hundreds of miles. He was offered a salary of $4,500, less than half what he was making as a private consultant. But he took the job because he thought it was an important and interesting challenge.

Over the next few years, he devised a novel way to run trains on electricity. "The design was both bold and risky. It involved concepts that were unproven outside a laboratory, and required equipment untried in large scale railroad operations," Robert C. Stewart wrote decades later for the Historic American Engineering Record, a federal project documenting significant engineering achievements.

Murray's system was the first time anywhere a mainline railroad was electrified successfully. Modern Amtrak trains between Boston and Washington are direct descendants of Murray's designs. It was a "trailblazing effort that set the standard for American railroads," Stewart wrote.

After completing the railroad work, Murray traveled to the Rockies for a vacation. But his mind resisted rest. "Perhaps it was the vastness of the outlook from the mountains that led me to take a broader view of the problems involved in the work I had left behind—problems that had been faced too closely in the daily routine to be seen in proper perspective," he later wrote.

He decided, staring out at the broad vistas, that the United States should connect its regional power grids into one giant grid. Neighboring cities and regions could share power, helping each other out. An inefficient and duplicative system, which made it needlessly expensive, could be eliminated. He called it a "Superpower System."

"In our profession, we have been building fast and building well, but not always in co-ordination," Murray wrote in a 1925 book. "There is virtue in our national motto—*e pluribus unum*." Out of many disparate power grids, Murray wanted to build one giant grid.

Murray persuaded Congress to appropriate $200,000 to fund a

study of a power grid that would join Boston to Washington. His final report draws similar conclusions to Insull's Lake County speech. Within a decade, all of the Northeast's power system could be connected to share power efficiently. The 558 existing small power plants could be replaced by 273 larger power plants. All of the wires and power plants would cost $1.11 billion, a staggering sum at the time. This was enough to build a dozen Hoover Dams and a dozen Empire State Buildings. But using a coordinated system, Murray wrote, would deliver $239 million in annual savings.

Murray's report included several colorful maps that traced out a new transmission network. Large power plants at the coal mines of Pennsylvania and dams along the upper reaches of the Delaware, Susquehanna, and Hudson Rivers would send electricity to cities. Within a few years, new wires would reach the Niagara Falls region and the St. Lawrence River.

An article in the *Chicago Tribune*, headlined "A Revolt Against Coal Tyranny," praised the idea of building large transmission lines to carry power. According to the article: "A system more or less on these lines is rapidly developing. It is no idle dream. It is a certainty of the future." It noted that power lines can carry electricity up to six hundred miles, and that California was considering a 1,100-mile line.

There was a reason the *Chicago Tribune* was so enthusiastic about burning coal far away and bringing electricity via wires. The first electricity in the city came from coal-burning dynamos in the basements of downtown buildings. The smoke was horrendous. "At times Adams Street for two blocks was like a huge trough filled with smoke. . . . The volume was so dense that one could almost lie on it," said R. W. Francis, the chief engineer of a local power company in the 1890s. A group of businessmen, calling themselves the Society for the Prevention of Smoke, railed against the plant. Some coal boosters tried to put a good face on the pollution. "Smoke is the incense burning on the altars of industry. It is beautiful to me," coal merchant W. P. Rend said at a speech at the Union League. "It shows that men are changing the merely potential forces of nature into articles of comfort for humanity." With the advent of long-distance lines, Chicago Edison could relocate power plants and their sooty emissions far from downtown.

Congress never acted on Murray's Superpower idea. He moved to South Carolina, where he designed and built a dam to contain the Saluda River, creating a record-setting, forty-one-mile-long man-made reservoir. The South Carolina legislature named it Lake Murray.

He and his wife, Ella, built a house nearby that they called Bohemia Manor. He designed a glass-sheathed studio for her to paint watercolors, and built a small dam on the property to create a fishing pond. It came to be called Little Murray Pond. He installed a small generator to make electricity from the water. For years, the dam was the only source of electricity in that rural part of the state. He used it to light up the house and make ice, which he gave away to neighbors.

As his Superpower report gathered dust, his idea of an interconnected power grid of larger power plants and connected utilities slowly became a reality. In 1923, Philadelphia Electric Company, Public Service Electric and Gas of New Jersey, and Pennsylvania Power and Light studied linking their systems. A report estimated that if two hundred miles of high-voltage transmission linked the systems, they would collectively need fewer power plants. Pennsylvania Power and Light needed the most power in the mornings, as industries ramped up. The Philadelphia and New Jersey companies had the biggest demand in the evening. By pooling, there were fewer idle hours and savings were substantial. By 1927, the three companies began to create what they called the "PNJ Interconnection." It was the "world's largest integrated, centrally controlled pool of electric power," a historian later noted. It took three years to complete a ring of high-voltage links. Delays were due to acquiring all the rights of way needed to cross northern New Jersey.

An even more impressive linkage occurred in the southeastern United States. In 1926, a drought in North Carolina led to a shortage of power in the state. North Carolina kept the lights on by borrowing power from Southern Power Company. But Southern didn't have a surplus of power and had to get excess from the Georgia Railway and Power Company, which in turn took power from the Alabama Power Company.

Alabama had excess power thanks to a giant hydroelectric power station at Muscle Shoals built by the government to provide power to a nitrate-fixation facility to make explosives for World War I. Begun in 1917, it was completed in November 1918—the month the war ended. The dam provided more power than was needed locally. But its excess power came in handy when the weather was dry in North Carolina.

Murray's idea of connecting disparate grids into one grid in order to use inexpensive electricity when it was available caught the imagination of many engineers. In a 1917 presentation to a meeting of the American Institute of Electrical Engineers, an Alabama Power Company official promised to light up the entire South with power from his company's dams. "Within the next five or ten years power from Muscle Shoals and other Alabama hydroelectric plants will be transmitted to Memphis, Tennessee, across Mississippi and even to New Orleans, Louisiana, a distance of approximately 300 miles," he said. "When this is done, seven states with a population in excess of 16,000,000 people will be served from what, due to interconnection, will be virtually one great system."

A century later Aaron Bloom, a supervisor at the National Renewable Energy Laboratory outside Denver, handed me a copy of Mitchell's paper. "We've been interconnecting this system to integrate renewables since the '20s," he says. "What they did in Georgia and Alabama to connect the southern system was about balancing hydro from different river basins and different time zones."

Bloom's office is located in the foothills of the Rockies, the same area where William Murray went for a break from work and developed the idea for Superpower. NREL is a campus of engineers, mathematicians, and scientists. Many of them are dressed in hiking boots and flannel shirts, as if they might need to go on a five-mile walk through the Rockies at any moment. Bloom dressed differently. He wore black wingtip shoes, dress slacks, and a white, pressed shirt that had a couple buttons undone roguishly. He looked as if he might be on his way to give a presentation to a roomful of Silicon Valley venture capitalists.

We sat in a conference room with a giant computer screen, probably thirty feet long and ten feet high, on one wall. From his laptop, Bloom pulled up maps of the giant Eastern Interconnection—the massive grid that stretches from eastern New Mexico to Quebec. It was William Murray's Superpower zone, but even larger.

A few months before we met, he had led a team that published a massive study on the eastern grid. They had created a giant computer model of the Eastern Interconnection to look ahead to 2026. They asked, What happens if the amount of wind and solar grow to be 30 percent of the power generation? Can the grid handle that much generation from resources that, unlike a natural gas plant, cannot always be dispatched?

"No one had ever run a model that way we ran it. Sixty thousand transmission lines. Six thousand generators. Every five minutes. There was no data set that existed, there was no tool that could run it that big. And there wasn't a computer to run it on," explained Bloom. So they built it, painstakingly. For every power generator, they needed to input all sorts of data. How much power could it provide? What was the minimum level it could run at?

Bloom and his co-workers fed all of this data into NREL's supercomputer, named Peregrine, a three-story machine that has its own building on the campus. Machines pump water through the computers to keep it cool. They direct the heated wastewater under nearby walkways to melt snow and keep them clear.

The first time they ran the data simulation, trying to determine if the grid could handle large amounts of renewable energy, it took the computer nineteen days to make all the calculations. Then the team started looking at it. A data visualization specialist named Kenny Gruchalla noticed something that didn't make sense. According to the computer, the sun was rising in Nebraska at the same time as Maine, 1,500 miles to the east. Solar farms were springing to life along the fictional Great Plains when the sun was just a predawn glimmer on the horizon. This was a problem.

"Shit. We did it wrong," Bloom said when Gruchalla pointed out the problem. The entire team working on the data stopped everything to solve out a basic question: "Why is the sun rising ahead of the sunrise?"

Josh Novacheck, an engineer a couple years out of college, was the first to figure it out. It was a data problem. A column had been sorted incorrectly. Getting the data fixed was easy, but it would require running the entire program again. He texted Bloom. "We need to talk," he wrote. Bloom was in San Diego, meeting with Energy Department officials who had funded the work. Bloom told his sponsors he needed a few more weeks to rerun the computer model.

The final results were mesmerizing. Bloom showed me on the giant screen what the power system could look like in May 2026. The scenario begins at night when power demand is low. There are giant blue circles—wind farms in Oklahoma, Iowa, and Michigan sending power into the Southeast and Mid-Atlantic. Up north, Quebec's hydro plants are active, as are natural gas plants in Florida. As the sun rises, solar plants spring to life in Florida. The state sends excess power northward. Later when the sun sets, that reverses and Florida imports power.

What was on the screen was a computer model of how half the country could be powered with large amounts of wind and solar energy. Regional grids trade power back and forth throughout the day. Power passes from one grid to the next like a relay race. Wind power from the Great Plains moves into the giant Midwest grid, which was sending power out to Manitoba and Ontario. At the same time, the Midwest sends power to the Mid-Atlantic states, which send it to the Southeast and to the New York grid operator and from there up into New England.

"We have put to bed the *is this possible question*," Bloom said, referring to the often-heard claim that adding a lot of renewables will destabilize the power grid. "We can do this." Thirty percent renewables was not a problem. "We can do this and it probably makes sense. The question is 50 percent and beyond."

The computer model on the screen assumed several giant direct current transmission lines existed, running from the middle of the country to the East. "We have really, really good resources in the middle of the country where no one lives and there is not a whole lot of transmission," said Paul Denholm, an energy scientist who worked with Bloom on the project and had been sitting quietly in the conference room. He worked on an earlier study from the laboratory called RE Futures. It examined a future of 80 percent renewable power and asked what impact it would have on power costs. The answer, surprisingly, was not a lot. Renewable energy costs were falling and integrating grids would eliminate waste. The laboratory released the study RE Futures in 2012 without a lot of fanfare because they worried about being accused of being wild-eyed activists.

One of the big lessons from the RE Futures study was that you need to build more wires to create a giant networked system. "In general, it found that if you restrict transmission, costs go up. That is fairly straightforward. I mean, wind is really cheap and it's going to stay really cheap," Denholm said.

Later, over lunch at the NREL cafeteria, where quinoa burgers and okra gumbo were the specials, Bloom explained that he sees a theme emerging in his work on power systems. The bigger the grid, the more stable and capable it is. Over the past few years, engineers' understanding and appreciation of networks has grown exponentially. After all, what is Google but a giant network of information? The internet itself functions because it is a vast and interconnected network. If one node fails, information is rerouted to prevent a wider failure.

The same holds true for the power grid. It was a lesson that southern power companies first discovered in the 1920s, when Memphis Light and Power, Birmingham Electric Company, and the Georgia Railway and Power Company began to connect to each other. These isolated grids connected to each other and the resulting network was more stable, resilient, and capable. They can share power and use their power plants more efficiently. Networking made all of them stronger and more reliable.

The operations of a 2026 power grid were all occurring inside Peregrine. All of the interconnections and new transmission to move wind and solar around were electronic figments of NREL scientists' imagination. The research was grounded in technical realities and regulatory common sense. They didn't ask the grid to do something impossible. "A lot of people who are pure academics, they have never been in a control room," Denholm said. "Yes, we are looking at aggressive scenarios. But we are firmly rooted in the realities of how the power system works."

Their computer models showed hundreds of gigawatts of new wind and solar farms, better transmission, and even some high-voltage direct current lines tying it all together. But who would build this grid? Where would the money come from? Who were the politicians who would get behind this vision? And would there be regulators who put national goals ahead of local interests? As Murray had wondered, Would we be able to build an *e pluribus unum* electrical system?

Over the years, many people had dreamed of such a grid. In 2006, the chief executive officer of American Electric Power, Michael G. Morris, reflected on an effort by his company years earlier. Between 1966 and 1986, American Electric Power, which owned dozens of power plants and several regional utilities, had built giant transmission lines across parts of seven states. The purpose of this giant grid was to move power from coal plants of record-breaking size. One of the first opened in 1973 a few miles downriver from Charleston, West Virginia. It burned thousands of tons of Appalachian coal daily, delivered on barges. The electricity could move on the wires toward Ohio and northern Indiana.

Morris noted how difficult it had been to take these steps and lamented how difficult it would be to build a network that could efficiently move power around from region to region. "Transmission remains trapped between federal and state regulatory regimes, slowing development of a truly, and much needed, national interstate grid," he said.

Michael Skelly was trying to correct this lack of a national grid. He was one of a handful of people to realize that wind energy was about to get inexpensive, although he and many others didn't foresee that prices for solar would plummet even further. But the good wind and solar resources were trapped in places like Guymon, Oklahoma. He had once put together a presentation with a map that showed that the counties in the United States with the best wind were also the counties with the smallest electrical wires. The only way to tap into this renewable energy was to build new wires.

"If you agree with the thesis that renewable energy is getting cheaper and cheaper and cheaper, which we agree with and have seen," Skelly said, "then the challenge becomes how do we get it to market."

3

Braulio Carrillo

For many years, Michael Skelly was directionless. Soon after his twenty-fifth birthday, he wrote a letter to his parents from Costa Rica, where he was wrapping up two years in the Peace Corps. "Time to start clipping fun articles about fun things to do with your life and sending them down my way."

He was born in Manchester, England, in 1961. His father, Tom Skelly, worked for Metropolitan Vickers, a company that made steam turbines and generators. Soon after the birth, Tom received a job offer from General Electric. He went ahead of the family, crossing the Atlantic on a Boeing 707 and settling in Erie, Pennsylvania. Dolores Skelly and her three children—the oldest was three, Michael was two, and there was a baby—followed, crossing the Atlantic on the SS *America*. The family had two more children in the United States.

They eventually settled in Roanoke, Virginia. Michael was an inquisitive child and stubborn. One day a neighbor called his mother to complain that he was often late delivering the newspaper. Michael would sit at the end of the block reading the paper, the neighbor reported. That night, his mother relayed he had received a complaint. "Well, I need to know what I'm delivering," he replied. Dolores laughs at the memory. "He never gave us trouble, but he could argue paint off a wall," she said.

Skelly was an active child. He fell in love with bicycling at an early age. In high school, he hiked up to McAfee Knob, a noted lookout spot on the Appalachian Trail outside Roanoke, to drink beer with friends. He was a good student and got accepted to Notre Dame. After graduating from college, unsure of his next step, he applied to the Peace Corps.

Skelly was sent to the small town of Golfito in southwestern Costa Rica. United Fruit Co. built the city, once called Banana City, in the early twentieth century as headquarters of its country operations. An extensive labor strike led the company to leave Golfito in 1985. A couple of months later, Skelly arrived in town.

The region had "very serious economic and social difficulties," noted the Peace Corps country head in a 1987 letter. There were few jobs and little bank credit. Skelly worked with an electricians cooperative to set up an inventory control system and a radio advertisement campaign. But he spent most of his time working with the town's fishermen. In letters home, he complained about not having enough to read and the challenge of developing a workable business model. "Gotta figure out a new way for these fishermen to sell their fish," he wrote. There was only one buyer in the capital and "he's got these guys over a barrel, always, as they say here, playing games with the prices." He worked on microcredit programs to help the fishermen start businesses to get their catch to nearby markets. "I didn't know much about fish or like running microcredit programs, but you sort of figure it out," he said.

One lesson that Skelly took from his time in Costa Rica was the importance and influence of money. In 1989, he decided that if businesses were central to the operations of the world, he should understand how they worked. He enrolled at Harvard Business School. As he neared finishing his degree, Skelly decided he wanted to have a more interesting job than any of his classmates. His wife has a slightly different memory. She said his objective was to find a job where he didn't have to wear a suit and tie to the interview.

A Harvard professor, aware of Skelly's desire to do something different, introduced him to Donald Perry. Perry was a biologist who had pioneered the study of life in the canopy of a rain forest by building a steel platform held aloft in trees by steel cables. Perry was tall and charismatic. An author described him as "the Jacques Cousteau of the tropical forest."

"Did you ever see the movie *Medicine Man*? It's a really terrible movie," said Skelly. "Sean Connery plays this biologist who is looking for the cure for cancer up in the canopy of the rain forest. And if he finds the cure, he gets to hang with this chick, the hot young biologist." Donald Perry was a real-life version of Connery's fictional character, Skelly said.

Perry's canopy held an irresistible attraction for people who wanted to see the rooftop of the rain forest. Besieged by requests from scientists and visiting journalists, he decided to make a tourist version of his contraption. He teamed up with investors to build an aerial tram in Costa Rica to take tourists into the canopy for an up-close view of the ecological diversity. The group needed someone to live in Costa Rica, secure permits, and oversee the construction. At his professor's suggestion, Skelly met with Perry and they talked about this project. Skelly wanted the job, and it wasn't long before Perry offered it to him. He didn't wear a tie to the interview.

Skelly had never built anything before, but that didn't bother Perry. Skelly spoke excellent Spanish and knew the country. He hadn't known anything about running microcredit programs, but had figured that out. He would need the same approach to the tram. What he didn't know about construction he would figure out.

It wasn't difficult to convince his wife, Anne, to relocate to Costa Rica. They both had an adventurous streak. She had also been in the

Peace Corps before they met. She had been assigned to Mali. To make up their minds about Costa Rica, the couple flew there and stayed at the base of Arenal, an active volcano. "Every hour, it was like cannons going off and there was hot lava," she recalled. "It was awesome." After a long weekend, Anne agreed to move there. They returned to the United States, packed their suitcases, and moved to San José, the capital, in 1992. Anne was eight months pregnant with their first child.

The job was challenging. The venture had about $80,000 in funding, which needed to be stretched for months. The group ran out of money for a while until a more significant investment could be lined up from a venture capital firm in Boston. Not only was the funding troublesome, the bureaucratic headaches were significant. Skelly was the group's country manager. "He is really the guy who pulled that whole project together," said Perry. "And he was very smart and very energetic."

Skelly loved the work. "In the morning, I would go out to the site, drive an hour to the rain forest, through this national park, run around doing construction stuff, then I would take a bath in the river, then hop in the car, come back to San José and do business stuff. That was the routine," he said. The thousand-acre site for the tram was adjacent to the Braulio Carrillo National Park, home to two- and three-toed sloths, long-tongued bats, howler monkeys, jaguars, silky anteaters, and tapirs.

He was happy. His friends in Costa Rica included José María Figueres, the son of José "Don Pepe" Figueres, who led the 1948 revolution that established democratic rule in the country and abolished the military. Skelly had met the younger Figueres while at Harvard and worked for him as a consultant while still in business school. While Skelly was building the tram, Figueres, thirty-nine, became the youngest president in Costa Rica's modern history. During his one term, he helped steer the country toward ecotourism and parks, now the nation's largest industry.

Skelly felt the tram could help save the rain forest, as well as provide jobs and economic development. But his motivation wasn't ecotourism. It was adventure. He had a job that was fun. There was a never-ending source of challenges and hurdles, puzzles that needed to be solved. He wasn't pushing papers. He was thirty-two years old and was trying to build something big in a remote Costa Rican rain forest right up next to a national park.

Nothing was simple. He needed to build a short gravel road from the main undivided highway to the tram. Perry recalled that they couldn't get the necessary permits, so they went ahead and built the road anyway, gambling that the government wouldn't be too upset. Skelly said he had obtained permission, or thought he had. The federal police arrived when the tree cutting began. They insisted that a different permit was needed to cut down each tree. Local prosecutors stopped work on the road and lodged a criminal charge of destruction of protected forests against Perry. A few months later, someone firebombed the court building with all the evidence and paperwork, presumably to wipe out government evidence against drug smugglers, not road builders. Prosecutors decided not to file a new case.

Perry's idea was to string up a tram through a mile of rain forest, allowing tourists to slowly travel through the canopy of trees in an open-air gondola. The slow-motion trip would last about forty minutes. Playing promoter to a visiting reporter, Skelly bragged that the tram would allow thousands of tourists "to be in the rain forest, to feel it, smell it and see its richness."

Skelly purchased a tram from a ski resort that was selling its old equipment. It was inexpensive to purchase, but hard to get into the rain forest. Each tower needed to be airlifted into place and then put on a twenty-five-ton foundation of poured concrete, hauled in one 110-pound bag at a time. Perry insisted that Skelly build the tram without knocking down any trees.

Skelly looked into hiring oxen to drag the towers into place, but rejected the plan as too destructive to the forest. So he started looking for a heavy-lift helicopter. He tried to secure one from Helicol, a company in Colombia, where left-wing guerrillas were blowing up electricity transmission towers. The country's military had contracted with Helicol to help rebuild the grid. As a result, the company had experience flying towers into remote areas. But he couldn't get a deal done. The U.S. Army in Panama had the hardware, but wasn't interested in lending it out.

At Harvard, Skelly had met several former Sandinista guerrillas-turned-political-leaders from Nicaragua. They were attending the Kennedy School of Government after losing an election. Skelly used to attend get-togethers at the Nicaraguans' apartment overlooking the Charles River. In search of a helicopter, Skelly flew to Managua and looked up his Harvard friend Salvador Mayorga.

Mayorga was happy to see Skelly and asked what he was doing in Nicaragua. Skelly responded that he needed a helicopter. In the 1970s, Mayorga had been a guerrilla and once had helped get a wounded Sandinista to a safe house. The wounded guerrilla was Colonel Manuel Salvatierra, who went on to become head of the Nicaraguan air force. Recounting this story, nearly three decades later, Skelly enjoys the way it turned the old-boys' Harvard network on its ear. This wasn't one Harvard alum picking up the phone to help another land a job at a Wall Street investment firm. It was a Harvard alum picking up the phone to arrange for a military helicopter to be used to move towers in a rain forest.

With Mayorga's introduction, Skelly got a meeting with Salvatierra. The men agreed on a price: $2,500 an hour, plus fuel and expenses. They shook hands. Skelly returned to Cost Rica and waited. Six weeks went by. The construction project was at a standstill while the towers sat idle at a port on the Pacific Ocean.

In the summer of 1994, Skelly returned to Managua and parked himself in the anteroom of Colonel Salvatierra's office near the Sandino International Airport. He felt like he was a character in a Gabriel

García Márquez novel. The receptionist was nice and brought him coffee. Patient and determined, he sat and waited. He was bored and stared out a small skylight, wondering if U.S. missiles once targeted where he was sitting.

He stayed there for an entire week before his obstinacy won out. The colonel realized Skelly wasn't going to leave until he got his helicopter. Salvatierra relented and told him the helicopter was his. There was one final delay: the colonel ordered that the helicopter's machine guns be unbolted. Skelly caught a ride home on the newly civilian aircraft, an Mi-17 built in the Soviet Union. "I wasn't going to pay unless I was in the helicopter," he said. Flying a foreign war vessel into Costa Rica's sovereign airspace required an act of congress. "We kind of skipped that part," Skelly said. He did manage to work out a deal with the local Costa Rican police. In the end, a joint Costa Rican police–Sandinista air force operation would move the towers into place.

When the first tower was flown in, the wire holding the tower wasn't long enough and the downdraft from the helicopter's blades mashed down the canopy they were trying to preserve and show off. "We knocked over a bunch of the forest," Skelly recalled. Perry was apoplectic. They made adjustments and installing the other towers went more smoothly.

By October 1994, the tram was inaugurated. Skelly had managed to overcome myriad financial, legal, and logistical problems. He had built a ski lift in the rain forest. Tourists could board an open-air gondola hung from a steel cable and travel through the trees, while a local guide pointed out flora and fauna.

The tram's financial backers liked what Skelly accomplished. They decided to build more, eyeing opportunities in Jamaica, St. Lucia, and along Costa Rica's Pacific coast. But Skelly wasn't interested in developing ecotourist theme parks in different places. He liked the challenge of building a tram, but not running it. By the time the first tickets were sold, Skelly had already checked out. He and Anne decided to return to the United States.

He was proud of what he had accomplished. But the tram's value to Skelly was as a calling card. This is what I can accomplish. With all its bureaucratic detours and unexpected headaches, the tram was good preparation for the second phase of Skelly's career.

———————

Skelly was looking around for a new opportunity when a friend introduced him to the chief executive of New World Power Corp., which owned, operated, and built renewable energy projects. Skelly accepted a job with the company. He was supposed to develop renewable energy projects, leasing the land, contracting for construction, and selling the future power. In reality, New World Power was in financial trouble. To conserve liquidity, it began to sell off assets such as a hydroelectric dam in Costa Rica and wind farms in California and the United Kingdom. Skelly was hired to be a builder, but the company was shedding assets. Still, he liked what he saw in renewables. When New World Power ran out of money in 1996, Skelly moved on.

In the summer of 1996, the family moved to Minneapolis, near where Anne had grown up. Skelly was about to turn thirty-five. Their third child was on the way. It was time to try being what he calls a "normal corporate person." He interviewed at Cargill, General Mills, and 3M. He didn't get any offers. He figured they could tell he didn't really want the job and wasn't cut out to work in a headquarters of a global company. Like an immune system that can identify something that doesn't belong and expel it, corporate America had rejected Skelly. The feeling was mutual.

Energía Global, a company building small energy projects in Central America, offered Skelly work as a consultant. Needing to make some money, he accepted. Within a few years, he was working full-time for them and his family had moved back to Costa Rica. He liked putting together energy projects. It was more challenging than even building a tram in the rain forest—and at the end of the project, he felt like he had accomplished something more meaningful than a tourist attraction.

At Energía Global, one of his first assignments was to examine a proposed wind farm called Tierras Morenas that was being offered for sale. He liked what he saw. The location was good: there were strong, consistent winds in the area and lots of land. And the size and ambition of the project appealed to him. It would be the largest wind farm in Latin America. But Energía Global was a small company and needed a partner if it was going to buy Tierras Morenas. Skelly heard that a wind company in Texas might be interested. He made a cold call and flew to Houston for a meeting with an investor named Michael Zilkha, who owned half of a company called International Wind.

Skelly pitched the project to Zilkha, who promised to look into it after a vacation in Costa Rica. Skelly arranged for Zilkha to have dinner with a former president and to take a tour on the aerial tram.

By the time the vacation was over, Zilkha decided he wanted to work with Skelly. "He was fun and smart and I liked the path he had taken," said Zilkha. Zilkha's wife and daughter liked him too. Michael Zilkha agreed to invest in Tierras Morenas. With Zilkha and his money on board, the project went forward. It would be Skelly's first real attempt to build a renewable energy project.

The plan was to erect thirty-two turbines along ridgelines where the vegetation had been stripped by cattle grazing. The towers were 130 feet tall and could produce, under the right conditions, about 750 kilowatts each. Each could power, on average, about three thousand Costa Rican households.

Skelly worked alongside a Canadian engineer named Rick Winsor, who was responsible for tidying up projects and readying them for construction. He was detail-oriented. "He would bring the ideas, but they weren't fully baked," said Winsor. "I would say 'here comes another one I need to fix.'" They argued often, but never got into a fight that couldn't be resolved.

Together, they developed, financed, and built Tierras Morenas. In August 1999, electricity began flowing from the turbines onto the Costa Rican power grid. By 2001, 4 percent of power in the country came from Tierras Morenas and two other smaller wind farms.

By then, Michael Zilkha had seen enough to know he wanted to continue working with Michael Skelly. Skelly had enjoyed renewable energy development and decided he wanted to pursue full-time work in the industry. He left Energía Global and moved to Texas.

———————————

Skelly's timing was good. At the end of the 1990s, the wind industry was in the last phases of a metamorphosis. It had gone through nearly two decades of debacles that had tarnished its reputation. It was about to become a legitimate power source. To the extent that the public thought about renewable power, it didn't take it that seriously. It was for dreamers who wore sandals. Power meant burning coal or natural gas, damming rivers, or splitting atoms.

At the time, when most people in the United States thought about renewable energy—if they thought about it at all—the most common image was of California's wind farms. Clustered around Palm Springs and mountain passes east of San Francisco and north of Los Angeles, these crowded fields of overgrown windmills perched on lattice towers were objects of derision. Most of the turbines didn't spin because they had broken down.

In the 1980s, there had been a wind rush in California's mountains where cool Pacific Ocean air flowed toward the state's warm inland valleys. There was no question these were windy areas. But the state did a poor job of designing incentives to boost the industry. It offered a tax credit for electrical production *capacity*. To get the tax credit, you had to install a wind turbine that could produce power. Later tax policy emphasized actual electrical *generation*, sensibly requiring the wind machines to do the job they were intended to do. What was built in the windy mountain passes was little more than tax shelters perched on spindly metal towers. Birds nested in the metal trusses of the lattice towers, turning each turbine into a spinning vertical Cuisinart that often was better at harming birds than generating electricity.

Toward the end of the decade, the wind industry tried to build a better wind turbine. A San Francisco company, Kenetech Corp., boasted it had created a new variable-speed wind turbine that overcame a long-standing problem of handling gusty wind. Able to speed up and slow down with the wind made the turbines more durable, so they didn't break down as often. This new technology was also better at capturing the power of higher-speed wind. Company officials said the new turbine could produce power at 5 cents per kilowatt-hour, comparable to fossil fuel generators. (A kilowatt hour is roughly equivalent to running a vacuum for an hour, or playing a PlayStation or Xbox for five hours.) In 1993, *Discover* magazine gave the new turbine its environmental innovation of the year award. "Unlike previous turbines, the 33M-VS is rigged to roll with the wind's punches. . . . The wind gust problems—the wear and tear and wasted energy—have all but blown away," the magazine gushed. The company had held an initial public offering a year earlier. The stock's value soared.

Kenetech's chief executive Gerald R. Alderson promised in 1993 that the company would be the "undisputed leader in wind-power technology" for at least the next five years. But the new turbine was flawed and so was the business plan. Within two years, the Kenetech board let Alderson go. Eighteen months later, the company filed for bankruptcy. The heavily promoted new turbines had mechanical problems, including blades that weren't properly secured and flew off. A bigger problem was a federal government ruling that invalidated contracts with California utilities. But just when the industry outlook appeared bleak, the situation began to turn.

Skelly saw it when he returned to Texas. The state had recently ruled that utilities could offer to sell their customers green, renewable power—and charge a premium for it. A month later, a large modern wind farm, using a new generation of turbines imported from

Denmark, began operation in Big Spring, Texas, a dusty town in West Texas where the eponymous spring no longer flowed. The project included the largest turbines ever installed in the United States: four giants manufactured by the Danish company Vestas that could generate 1.65 megawatts each. These were more than twice as large as the turbines installed in Tierras Morenas and each could power 1,200 U.S. households. The Vestas turbines were physically larger than older wind catchers. They were twice as large as what was in California—the blades were 66 meters compared to 33 meters for Kenetech's hyped machine. Their size made the machines more cost-effective, as they could produce power for less. The Big Spring turbines, erected in the late 1990s, marked the beginning of the modern wind industry in the United States.

These turbines—indeed, every new turbine constructed in the United States—were built on monopoles. These skinny towers replaced the slender pyramid lattice designs. There were no longer any places for birds to roost. The new larger blades spun slower, also reducing bird fatalities. These turbines introduced new high-tech touches that became standard: weather meters that tracked the wind and automatically turned the turbine and each individual blade to maximize output. Wind turbines were being built taller, to capture more of the strong and steady wind at higher elevations.

The new poles and taller turbines weren't what caught Skelly's attention. What caught his interest—and many others who subsequently got into the wind business—was what soon happened to the cost of generating wind power. It fell precipitously, just as the price of solar power would fall a decade later.

The first indication of these falling prices occurred on April 1, 1999, although it wasn't an April Fools' joke. That day, the Waco, Texas, mayor's office received a fax from the headquarters of Texas Utilities, the giant utility that provided electricity to Waco, Dallas, and a large portion of the rest of Texas. The fax detailed a press conference scheduled for the next week. Texas Utilities planned to roll out a new, optional power plan called "TU Renew." Residents and businesses could choose to have some of their electricity come from new wind turbines near Big Spring. Texas Utilities had conducted a poll the year before that found 96 percent of respondents wanted it to invest in more renewable energy, and political leaders in Austin were considering new rules to require the company to incorporate wind energy. TU Renew was a way to figure out how to make a profit off these wind turbines.

The invitations sent out to local media included a hat with a logo and the phrase "Together, we can generate change." A red-and-green pinwheel, the kind you blow into to turn, was tucked into the envelope. When the program was announced, officials emphasized the size of the turbines. One person pointed out that the turbines' blades were nearly the size of the twenty-two-story Alico building, a 1910 stone skyscraper that still dominates Waco's skyline. Corporate fact sheets emphasized that the lights wouldn't go off even if the wind stopped blowing.

Dan Farell, an executive with Texas Utilities, spoke at the Waco press conference. He talked about a recent meeting the company had with customers from across the state. "Those customers said they'd be willing to pay the extra costs of renewable energy if we made that option available to them. So that's what we're doing today," he said.

Farell was referring to a series of remarkable meetings that utilities had been required to hold by the state of Texas. As part of a planning process, utilities had selected a small group of customers to come to a hotel for a weekend. They received a crash course in how the state generated electricity, as well as the trade-offs and costs of choices. The results surprised everyone, from the academics contracted to run the weekends to utility executives and also to the state's top regulator. The customers said they liked predictable costs. And even though they were told it would cost more, a majority of attendees said they wanted more renewable energy. They liked the idea of cleaner energy sources. These meetings weren't taking place in wealthy, liberal enclaves. They were happening in Corpus Christi, Abilene, Beaumont, and Amarillo. "I think it opened their eyes—here were ordinary people interested in renewables," recalled Dennis Thomas, one of the people who helped run the polls.

Customers said they would pay a little more for renewable energy. Texas Utilities decided to find out if their actions matched their words. Their test would be in Waco. The midsize city, not the state, had jurisdiction over its own power rates. The rate Texas Utilities rolled out in April was quietly exceptional: 4 cents for every kilowatt hour used. This was about the cost of generating power by burning natural gas in a modern power plant.

A few years earlier, Kenetech had boldly promised 5 cents and people scoffed. It was an unrealistic dream. Now Texas Utilities was offering wind power even cheaper. The cost of generating electricity from wind had begun to fall as the wind turbine technology improved. Renewables were no longer the province of dreamers, people who wanted to divorce themselves from fossil fuels with a few solar panels and wind turbines. The price of wind was getting closer to being competitive with coal and nuclear and natural gas.

"That was a wakeup call," remembered Skelly. "Nobody could believe it." What happened next, in the summer of 1999, stunned everyone paying attention. The city of Austin, which owned and operated its own power company, decided to offer its residents renewable energy. A liberal university town with a burgeoning high-tech industry, the city wanted to burnish its green credentials. It put out a call to companies and received a dozen offers. Enron offered wind at 4 cents per kilowatt hour. But when the Enron executives met with Austin officials, they learned they weren't the low bidder. They had been undercut by Texas Wind Power Co., which offered 20 megawatts of wind for 2.85 cents, according to Mark Kapner, the utility's manager for renewable energy. It was coming from a new wind farm in rural Upton County, in arid West Texas. Upton County was a favorite target of renewable developers. A couple of years earlier, the Texas Renewable Energy Industries

Association had distributed a brochure. If Upton County were covered in solar panels "it would provide all the electricity presently consumed by Texas each year," it said, making no mention of the three thousand inhabitants who would need to be relocated.

In July, the utility asked the City Council for approval to enter into a contract with Texas Wind Power. The resolution disclosed the size of the contract both in terms of megawatts and dollar amounts. From there, it was easy for people in the power industry to figure out the price.

———

Wind was getting cheaper to generate, although in most places it was still more expensive than natural gas, the next cheapest source of power. But there is a critical difference between wind and gas. To run a natural gas plant, you have to buy fuel constantly. And fuel costs can be volatile. With wind, the fuel is free. There are still significant costs for the turbines and construction, maintenance and operations, and so forth. But these can all be figured out on a spreadsheet. And the wind developers had a good idea of how many hours in a year a turbine's blades would be turning, spinning magnets, and generating electricity. So building wind enabled a power company to do something unusual in the energy industry: offer a fixed price contract.

Austin Energy sold its renewable power to residents and businesses at 1.7 cents for a kilowatt hour of power, after a hefty subsidy approved by the City Council. There was also a generous federal subsidy to wind companies designed to spur the growth of a new, domestic source of power.

A few residents signed up. Kapner said it was a "slow and gradual" process and the city used bill inserts to promote it. But the city's industrial users didn't have to be convinced. The city's microchip manufacturers, such as Motorola and Advanced Micro Devices, were eager to sign up. Austin Energy was offering to lock in power prices. They could either buy power for 1.4 cents for every kilowatt hour, a price that was subject to change anytime if prices rose or fell, or they could sign up for the new "Green Choice" and lock in their power prices at 1.7 cents for a decade. The choice was easy.

Then natural gas prices rose throughout 2000, and spiked in December. The cost of 1,000 cubic feet of natural gas at the beginning of 2000 was $2.16; it hit $10.49 in late December. Austin Energy adjusted its electrical prices upward. From 1.4 cents, it rose to 1.6 cents in August, then 2.2 cents in November—at which point it was cheaper to buy wind power. By February, it raised its prices again to 2.7 cents, just shy of the unsubsidized cost of wind power. Residents and companies that had signed up for green power were paying *less* than people getting power from the utility's typical mix of coal, nuclear, and natural gas.

Watching this, Skelly was convinced that the wind industry was about to take off. He was anxious to be a part of it. There was one other critical development. Texas was deregulating its energy industry. Gone were the companies with monopolies over certain geographic areas that generated power and delivered it on wires, earning a state-guaranteed return on their investment. In January 1999, a bill had been introduced into the legislature to create a competitive market for power generation. The bill also included a requirement to add 2,000 megawatts of renewable energy—doubling the amount of wind and solar in the entire United States.

Governor George W. Bush liked the idea of a competitive market and was willing to go along with the renewable energy requirement. Some of his political supporters, including Enron chairman Ken Lay, pushed him to embrace wind. Lay was trying to drum up business for Enron's recently acquired turbine manufacturing division. Conservative ranchers in West Texas, an important constituency of the state's Republican Party, also supported the idea. Building wind farms was good economic development for the rural parts of the state left behind by the economic growth in the cities.

In June 1999, Bush signed electric deregulation into law. The bill included the 2,000-megawatt requirement. The Lone Star State by itself would double the size of the national renewable energy industry by 2009. At the time, it seemed like an unrealistic bit of Texas braggadocio. It would turn out to be ridiculously low. Texas hit the 2,000-megawatt requirement and kept going. The state legislature increased the mandate to 5,880 megawatts by 2015. That would also, in time, prove too conservative.

In 2000, the American Wind Energy Association gave Texas state senator David Sibley an award at its annual conference because of his role in the passage of the Texas renewable mandate. "Wind energy is here to play and stay, and can bolster our energy security and competitiveness," said Sibley, an oral surgeon who gave up his practice due to a bad back and became a Republican politician. The conference was held in Palm Springs, California, and attendees celebrated what they hoped was the beginnings of a new industry. Outside of town, the dregs of the old wind industry were visible: broken-down wind turbines that few bothered to salvage for parts.

———————————

Skelly moved to Houston to work in this new and rapidly growing wind industry. He was hired to be the chief development officer at International Wind, the company he had convinced to invest in Tierras Morenas. Everyone around him in Houston's energy business seemed to be going to work for Enron. His friends and Harvard classmates were becoming "dotcom gazillionaires."

He was setting off in a different direction. "I was in a business that some called an unholy marriage of hippies and accountants," he recalled later. "But we believed in the promise that maybe this technology would get better, that maybe the country would start taking clean energy a little bit more seriously, and we happened to start doing it at a time when the wind industry really took off."

4

Johnny Rotten

As the year 2000 began, renewable energy was on the outside looking in. There were some new wind farms being developed, and a smattering of interest from large companies. But it was still regarded widely as a fanciful gimmick, a cute diversion from the grown-up business of burning fossil fuels. Outside of the governors' mansions in Austin and Des Moines, few politicians gave it much attention.

Even the U.S. government wasn't sure how committed it was. A year earlier, Congress allowed the subsidy for wind power to lapse for several months. And it had only extended it for two years, barely enough time to get a wind farm platted and built. Its brightest star, Kenetech, had taken less than three years in the round-trip from an initial stock offering on the New York Stock Exchange to bankruptcy court.

So who would want to invest money into this fledgling industry?

The answer was not many people. There were a few true believers who saw renewables as a way to break free from fossil fuels and their environmental toll. A regional utility called Florida Power & Light, based near Palm Beach, Florida, was interested. It stuck to gas, nukes, and coal on its home turf, but it was beginning to branch out into wind in Texas, Iowa, and the Pacific Northwest.

And there was the wealthy Zilkha family of Houston, by way of London and Baghdad. Michael Zilkha was a bit of a misfit capitalist. He could speak the jargon of calculated returns on investment, but preferred to discuss art and music. He had a philosophical bent. His conversations could veer in unexpected directions.

Michael Zilkha came from an Iraqi-Jewish family whose business acumen had been apparent for years. The patriarch, Khedouri Zilkha, Michael's grandfather, founded a small bank in Baghdad in 1899. The Zilkhas were part of a Jewish population that could trace its roots in the city back to Nebuchadnezzar II—six centuries before the birth of Christ. They lived in a community that revolved around arranged marriages and business deals sealed over strong coffee and water pipes.

In 1927, the Zilkhas fled Baghdad in the face of rising corruption and extortions of Jewish families. The family spread out to Beirut, Alexandria, London, and New York to create a financial and foreign exchange business that soon spanned five continents. When World War II broke out, part of the family fled the Middle East for the safety of the Upper West Side of Manhattan. In the decade after the war, the Zilkhas suffered significant financial setbacks when the governments of Iraq, Syria, and Egypt expropriated their businesses.

Three brothers eventually ran the family banking business, known as Zilkha & Sons. The youngest, Selim, Michael's father, grew tired of finance. He decided that he "would rather borrow than lend," according to Selim's brother Ezra in a 1999 family memoir titled *From Baghdad to Boardrooms*. He wanted to create something, a tangible business, rather than move money around the world to facilitate others' dealings.

Impressed by a French retail chain that catered to mothers and infants, he decided to create similar stores in England that he named Mothercare. The stores catered to mothers and mothers-to-be who wanted to be waited on by trained staff who sold them everything from bouncing cradles to plastic building blocks. The company was ahead of its time. Selim Zilkha bought an early IBM mainframe to manage inventory. A decade later, the Mothercare empire included 143 stores in four European countries. *Time* magazine named him "The Baby King."

In January 1982, Mothercare merged into another company and Selim sold his interest. He relocated to the United States. His son, Michael, was already living in New York. Michael Zilkha wasn't interested in banking or retail. At the time, he was interested in punk and disco music.

Michael Zilkha was born in 1954. When he was six, his parents' marriage ended when his mother, Diane, had an affair with a prominent member of the British Parliament, Lord Lever. They eventually wed, a marriage that lasted for three decades until his death. Michael attended boarding school at Westminster, a posh academy in London. From there, he went to Oxford University, where he studied economics, then switched to French literature. After graduation in 1975 he moved to New York City. He was twenty years old. He wrote theater reviews for *The Village Voice*, an alternative weekly, where he met and fell in love with another *Voice* contributor, a Harvard dropout named Cristina Monet, who wore Adidas sneakers with dresses. She also sang. A couple of years after they met, Michael Zilkha set up a label called ZE Records. The first album featured Cristina, in an attempt to bring together punk and disco, two musical forms coursing through New York's veins at the time. "I just thought that you could unite the two and make dance records that survived repeated listening," he said in 1984. *New York* magazine called the couple "Les Enfants Terribles de Rock 'n' Roll."

It wouldn't have been a surprise if ZE Records had been an artistic and a financial failure, a toy for the scion of a wealthy family. But that wouldn't be a fair appraisal. The label didn't make any money, but it left a cultural mark. The influential British rock magazine

Melody Maker selected its first record, "Disco Clone," as the single of the week. Another British magazine called it "the most fashionable label in the world." Zilkha had a wide and varied taste in music. He helped launch Kid Creole and the Coconuts, a disco-Latin music fusion experiment. In the 1980s, the Waitresses' "I Know What Boys Like" and "Christmas Wrapping" were staples on rock radio stations. The influential punk duo Suicide and Was (Not Was) were on the label's roster.

"When Michael Zilkha started out he was just seen as this rich kid. He was universally regarded as a joke. But time and again, he brought his projects home," said Paul McGuinness, a friend of Zilkha's who was managing the up-and-coming Irish band U2, in the *New York* article.

By 1981, the thrill of producing interesting but commercially unsuccessful music was wearing on Zilkha. "I don't give a damn about having the hippest record label in the world," he told a music journalist at the time. "I'd much rather have the hits." The hits never arrived.

Zilkha folded ZE Records in 1984. He had lived in New York for nearly a decade and had a marvelous time in the last days of disco. He had produced memorable music, and had even helped finance a forgettable, blood-soaked slasher movie. But he had also spent a good amount of his trust fund. He relocated to Houston to work with his father. The enfant terrible became an oil executive.

Building an oil company was a contrarian bet. Oil prices were falling. A prolonged glut of crude left the U.S. energy industry in crisis and Zilkha's new hometown shell-shocked. Houston was full of see-through skyscrapers: brand-new office towers with empty floors. The city would shed 150,000 jobs between 1983 and 1986, many of them in the oil industry. It was a terrible time to be an out-of-work petroleum engineer, but a great time to be starting up a new oil company and hiring.

Maybe it was because he was older, or working with his father, but Michael Zilkha turned out to be a much better oil executive than record company executive. He didn't know geology or engineering, but he was smart and a quick study. And he saw opportunities that others missed.

Before long, Zilkha Energy Co. began exploring for oil and natural gas in the Gulf of Mexico using computers. In several years, this practice would become standard, but at the time Zilkha Energy was leading the way. The industry used seismic waves to map underground reservoirs of oil and gas. These surveys were expensive to purchase, and due to their size, expensive to store digitally.

The Zilkhas bought all the seismic data they could and used it to figure out where to drill. They spent $1 million in 1992 on a terabyte of data storage, about the capacity of four Apple iPhones. The seismic data was housed in large cabinets that Michael Zilkha remembered as resembling port-a-potties. In 1996, the company spent more than $1 million on an Auspex NS7000 computer server. It was the second of these powerful machines to arrive in Houston; the first went to NASA. A year later, Zilkha Energy bought another Auspex.

Taking their cues from the data, Zilkha Energy bid on offshore leases. The list of most active bidders contained names of many large companies: Shell Offshore Inc., Chevron USA Inc., Conoco Inc., Exxon Corp. By the mid-1990s, another name was sandwiched amid the big boys: Zilkha Energy Co. At times, it was the most active bidder. They avoided attention. The company never issued a single press release.

They were the first company in Houston to marry big data with oil exploration. And it worked. A few years after its founding, Zilkha Energy made a discovery in two-thirds of the wells it drilled in the Gulf of Mexico. Other companies wanted this magic touch. In 1997, Sonat Inc. offered $1.04 billion for Zilkha Energy Co. The Zilkhas tapped Goldman Sachs, the Wall Street firm that had handled the family business, to conduct the sale.

The decision to sell was easy. The Zilkhas had built up a big advantage over competitors by buying more data and having the computing power to crunch it all. But Michael Zilkha saw that the cost of data and computing power was falling. Their competitive advantage wouldn't last long. And the Zilkhas had begun looking at wind power.

Zilkha Energy had a long-standing deal to sell the natural gas it produced to Enron. To keep abreast of his buyer, Selim Zilkha read Enron's securities filings habitually. These were inscrutable documents filled with wishful market-to-market accounting and ultimately outright deceptions. At the time that hadn't become apparent to even the most suspicious Wall Street investors.

Selim Zilkha read them and remarked that of all Enron's sprawling tentacles, the only thing he was jealous of was its renewable energy business. In early 1997, Enron had single-handedly rescued the fledgling wind industry by buying turbine manufacturer Zond Corp. This intrigued the elder Zilkha.

When the Zilkhas sold their company to Sonat, the father and son had become wealthier than ever but also had tired of fossil fuels. Selim Zilkha was seventy years old. He had built and sold two hugely successful businesses. His son was worried that without a business to run, his father would fade. Michael Zilkha's solution was to create a new enterprise that would engage Selim and allow both of them to move beyond fossil fuels. Michael remembered his father mentioning his interest in Enron's renewable energy assets. Why not start a new renewable energy business? The deal to sell Zilkha Energy closed in February 1998. "We started our wind business immediately," says Michael Zilkha. "The next day."

The first purchase was a 50 percent stake in International Wind. The investment didn't turn out to be what they had hoped. Michael and Selim Zilkha wanted to use the company to make increasingly large investments in wind power. After investing in Tierras Morenas, Skelly had joined International Wind as the chief development officer. Michael Zilkha liked Skelly and wanted to invest in him. Zilkha was frustrated by the slow pace of development. So was Skelly.

George Hardie had founded International Wind. A former tennis professional, he got into real estate and then wind when his sports career fizzled out. He had managed to become the eightieth ranked player in the world. He was good enough to play in sixteen Grand Slam tournaments, but not good enough to be anything more than early-round road kill for the world's best players. By 2000, Hardie had run out of money and didn't have access to more. His investors were friends. None had particularly deep pockets. Since the partnership was a fifty-fifty split, the Zilkhas and Hardie needed to put in equal amounts of money to fund development. That wasn't a problem for the Zilkhas, but raising money was a constant headache for Hardie.

Skelly wanted to work for a company that built something. Hardie's constant scratching around for funds meant that Skelly didn't have much to do. In 2000, he flew to Houston and informed Michael Zilkha he was going to quit. He had an offer from Texas Wind Power Co., which had built the wind farm that sold power at such a shockingly low price to Austin Energy.

To Skelly's surprise, Michael Zilkha told him he was frustrated too and didn't want to lose him. Michael Zilkha credits part of his early success in the oil business to hiring a young petroleum engineer he and his father trusted to handle finding the oil. Now he wanted someone like that he could trust to handle wind farm development. Zilkha believed Skelly would be that guy.

Zilkha wanted to step up his investment in wind, as long as Skelly would stick around and run the company. Skelly agreed, so Zilkha bought out Hardie for another $3 million and change, gaining control of the company. He moved its headquarters from Dallas to downtown Houston and renamed it Zilkha Renewable Energy. Michael Skelly moved his family again, the fourth move in five years.

They settled in West University, a Houston neighborhood with large trees, not far from downtown. Skelly commuted to work on a green beater mountain bike, dependable and not flashy enough to attract thieves.

Skelly was energized because the company now had money to spend. Mark Haller, a veteran of the wind business since 1981 and the third employee at International Wind, says it was the first time he could remember a wind company that was actually well-financed. He didn't have to scrimp and save. "I was glad to see that someone had some cash. That was an opportunity you didn't walk away from. I had spent fifteen years in the desert looking for water," he said.

The company began to hire new employees, a mix of young idealists and energy industry veterans who had spent years building conventional power plants. For all the youthful energy inside Zilkha Renewable, the outside world didn't look promising. As 2000 began, a year with its millennial turnover, the world of energy looked very much like it had for years.

The largest source of energy was oil, and most of that went into gasoline. On average, Americans annually consumed 450 gallons of gasoline and paid $1.56 per gallon. At the beginning of the year, General Motors canceled its EV1 electric car program, ending a short-lived experiment. But there were signs that the new century would also bring change. In the late summer of 2000, Toyota introduced a new car, the five-seater Prius, which could get an unheard-of 50 miles per gallon. It had both batteries and an internal combustion engine that ran on gasoline.

Old-fashioned electric utilities delivered the vast majority of power, much as they had in the first few years of the industry after Thomas Edison had brought online the first coal-fired power plant. Power plants burned one billion tons of coal to generate more than half of the electricity used in the country. One hundred and four nuclear power plants provided another 20 percent of the power, a steady output but one that was unlikely to change much. The last new order for a nuke in the United States had been in 1978.

In 2000, renewable energy provided 10 percent of the power in the United States, according to the Energy Department terminology. The majority of this came, 274.6 billion kilowatt hours, from decades-old hydroelectric dams. Of the remainder, burning wood scraps and garbage generated 60 billion kilowatt hours of power. Wind and solar combined: 5.7 billion kilowatt hours. By way of comparison, coal provided 1.96 trillion kilowatt hours.

The same year, the official government forecast for the next twenty years called for "continued growth and reliance on the three major fossil fuels—petroleum, natural gas and coal." The future for renewables? There would be a "modest expansion." The federal government wasn't helping. A wind subsidy expired in mid-1999, and lapsed for five months before Congress reinstated it. Over the next four years, Congress would allow the subsidy to sunset three times, only to reinstate it later.

In May 2000, Michael Zilkha traveled to Palm Springs for the annual gathering of the American Wind Energy Association, a trade group of wind developers and turbine manufacturers. The convention center was tawdry, he remembered. Compared to the sprawling oil gatherings of Houston, which fill up entire convention centers and then spill out into the parking lot, the wind gathering felt quaint. Michael Zilkha said the industry felt "somewhat messianic" and fragile. He thought it could fold up at any moment and disappear. There were only a few hundred people there. Selim Zilkha accompanied his son to Palm Springs. At one point, walking the floor, Selim turned to Michael and wondered aloud: "How are all these people earning a living?" Michael didn't have an answer. He had convinced his father to spend $6 million on a wind company. Now he needed to figure out how to make it work. He had a sinking feeling that the people at the conference were there to save the planet, not make money.

Still, the industry had grown quite a lot over the preceding decade. Randall Swisher remembered attending his first conference in 1990 as the newly hired head of American Wind Energy Association. It was held in a hotel with frayed carpets and pipes that clanged, located in an unglamorous neighborhood in San Francisco. He counted eight exhibits on tabletops. There were at least as many men with ponytails as there were men wearing suits. "There was no real business going on," he said.

At Michael Zilkha's first wind conference, everyone seemed to be talking about Sonny Bono. The entertainer had recently been elected mayor of Palm Springs, home to four thousand aging turbines. One of his top priorities was to tear down the turbines, in a bid to make the city more attractive to tourists. People at the conference discussed how to stop him.

The Zilkhas were something new in the wind industry. They understood finance and energy markets. They came from the fossil fuel industry, and in the eyes of bankers they had financial credibility that other renewable energy developers lacked. They also brought a large checkbook. They put a lot of money into projects, leasing up land and putting earnest money down to reserve turbines. "We never had more than about $100 million exposed in the wind business, at the most. So it was manageable," Michael Zilkha recalled. For many of the developers who had tried to create a wind business in the 1990s, it was an unimaginably large amount.

Zilkha Renewable Energy leased space in the same Houston high-rise office building where the oil company had been located. It was a twenty-three-story Art Deco masterpiece, built immediately after World War II, with marble floors and slow elevators. It stood out in a downtown filled with glass-sheathed skyscrapers. Michael Zilkha owned an enormous collection of rock 'n' roll photographs, and he decided to decorate the company's new offices with them. A giant black-and-white close-up of a glowering Johnny Rotten from the Sex Pistols hung in a conference room. Sometimes people from the music business passing through Houston who knew Zilkha in his New York days would stop by the office to look at the photographs. One worker swore he saw Debbie Harry in the office.

The new company struggled at first and so did Skelly. Wind development required the mastery of a hundred details: finding the right location with good wind, negotiating with utilities, and securing access to the power grid. Landowners were reluctant to take a chance on planting giant turbines amidst their crops.

In 2000, he attended a meeting in San Angelo of the Texas Renewable Energy Industries Association. There was one landowner there, Pat McDowell, a longtime executive with the aerospace firm Goodrich whose family owned seven thousand acres in the Texas panhandle. A self-described state politics junkie, he had followed the pro-wind changes and hoped that the turbines might provide some diversification to his family's ranch income.

He talked to Skelly at the meeting, and then bugged him to visit the ranch house, a ninety-minute drive east of Amarillo. McDowell said of Skelly: "He was odd and eccentric, but he wasn't crooked." After a little research, McDowell realized the money backing Skelly was real. Skelly installed a couple of towers to measure the wind speed at the height of a turbine. McDowell worried that when he introduced Skelly to the local county executives, they wouldn't trust him. Skelly didn't exactly blend in with rural Texas. "He was this tall, gangly Harvard graduate with these big wild ideas. He was not your typical Texas panhandle dude," said the rancher.

But Skelly won over the county commissioners with his earnestness and passion. Eventually, McDowell and Skelly met in the back room of the ranch house and plotted a wind farm for an exposed ridgeline above a canyon. It would be Skelly's first big wind farm and the first that Zilkha Renewable developed on its own.

Two weeks before proposals to the Amarillo-based utility, Southwestern Public Service, were due, Skelly called McDowell. "I have some bad news," he said. The project was dead. "I'm sorry."

Skelly had planned to feed the power from sixty-odd turbines into a transmission line that ran through the ranch and westward toward Amarillo. After he had paid McDowell several thousand dollars in down payments and had spent considerable time on the project, Skelly learned that Southwestern Public Service didn't own the line.

To the naked eye, the line looked continuous. But about ten miles west of the proposed wind farm, the line's ownership changed. The same large lattice towers held a triad of black wires carrying 115,000 volts. But on McDowell's ranch, the line belonged to AEP Texas, a utility (its parent company was American Electric Power) that served customers in a north–south band from the Texas panhandle down into the Rio Grande Valley. It was a different company.

Southwestern Public Service had put out a bid for wind power to be built in its region, serving its customers and using its transmission lines. AEP had no real incentive to help out. Even if it did, it could take months to hammer out connection agreements. If the McDowell Ranch had been ten miles to the west, there wouldn't have been a problem. But as it stood, it might as well have been on the dark side of the moon.

Skelly had found a good windy spot, but there was a problem with the wires. It was his first up-front view of the balkanized world of utilities and power lines, and how complicated developing renewables could be.

———————

Skelly and Michael Zilkha decided to add an operations manager, in part to make sure these kinds of mistakes didn't happen again. In the spring of 2001, they hired Rick Winsor, the Canadian engineer Skelly had worked with on Tierras Morenas. It was Skelly's job to dream up new wind farms out of the blue, looking at maps of wind strength and envisioning wind farms. Winsor made sure there were no hidden problems, snafus that would generate friction and slow down development. "We called ourselves the yin and the yang. He was the entrepreneur. He was the guy who had vision and great ideas. Some of which I thought were crazy," Winsor said. "I was the guy who tried to take the insanity and try to make it into reality."

Skelly liked to hire young, inexperienced people who were full of a passionate belief that they could change the world. Winsor hired engineers to make sure everything worked as promised. The two men fought. "Rick Winsor was very, very systematic and Michael Skelly was more laissez-faire and more of a dreamer," said Michael Zilkha.

On most Mondays, Zilkha would take Skelly and Winsor to lunch in the tunnels that run under downtown Houston, grabbing a salad or fast food. He would try to get them to work out their differences, so the conflicts didn't explode at the office. "At one point, I threatened to send them to my analyst," said Zilkha. "I felt they needed couples therapy."

The work was complex, but had one large benefit. They weren't gambling on untested technology. Wind power works. It had for more than a millennium. A traveler passing through eastern Iran, in AD 947, described a city where windmills were used to raise water to irrigate gardens. In 1887 and 1888, an inventor named Charles Brush built a wind generator in the backyard of a stately Cleveland home. It looked a bit like a giant Japanese oil-paper umbrella turned on its side. It was the first successful machine to turn wind into electricity. Remarkably, it operated for a dozen years, lighting up Brush's home until he abandoned it, as the energy historian Robert W. Righter noted, "for the convenience of using centrally produced electricity."

Skelly and the rest of the Zilkha Renewable team knew that the new wind turbines, sold by Enron and a couple of Danish companies, worked. The question was finding the right spot and keeping costs down. A strategy emerged: find the best windy locations that were near transmission lines, preferably in parts of the country with relatively high power costs. Then lease up as much land as possible and begin the often drawn-out permitting process.

Of course, this development work cost a lot of money. Every month, someone needed to make a cash call on Michael Zilkha. This task usually fell to either company chief executive Joe Romano, a longtime associate of the Zilkhas who had worked for the family's oil company, or Jayshree Desai, a young financial officer the Zilkhas had hired after she left Enron. Sometimes, the cash calls were weekly. Michael Zilkha usually wired the money from a family account. Sometimes, he would pull out a checkbook and, sitting at a desk in the office like everyone else, he would write a check. He made a show of writing a check every once in a while for a specific reason. "It is very, very important that people knew it was not some third-party money," he explained. It was the Zilkhas' money that was being spent.

One place that had wind, transmission, and good prices was southwestern Pennsylvania, where the rolling terrain created strong winds atop the hills. In 2001, Zilkha Renewable partnered with Atlantic Renewable Energy to build two small wind farms, one in Fayette County and the other in adjacent Somerset County. They bought the newest, most technologically advanced turbines from Enron, large 1.5-megawatt generators. Nicholas Humber, head of Enron's commercial wind sales, sold Zilkha Renewable the turbines and helped assemble both projects.

On September 11, 2001, Humber boarded a plane in Boston headed for Los Angeles to try to close another turbine deal. His plane became the first to crash into the World Trade Center. When word of the attack reached passengers on United Airlines Flight 93, they fought back against the hijackers. Bound for Washington, D.C., the plane crashed in Somerset County, seven miles northeast of the new Zilkha Renewable wind farm.

A month later, the inauguration of the wind farm turned into a memorial for Humber. A group of schoolchildren wearing T-shirts that said "wind is cool" sang songs. Skelly spoke, praising Humber for having spent most of his working life chipping away at America's energy dependence. "Wind power is patriotic," he said. "Because it's our resource."

The company placed a plaque honoring Humber at the base of

one of the turbines. When excavating the foundation for that turbine, workers had found an unmarked coal mine, Mark Haller said. Zilkha Renewable purchased extra concrete to seal it up.

At the company's other Pennsylvania wind farm, in Fayette County, Zilkha Renewable faced a noise problem. It had put the string of ten turbines atop a ridge. The homeowners in the valley below complained about the noise when the wind blew.

Michael Zilkha pulled out his checkbook. Zilkha Renewable paid for the homes to get high-density roofing, insulation, and new windows to block the building-penetrating low frequencies that a large wind turbine can generate. A couple of homeowners asked to be bought out. Zilkha agreed.

The company's ambitions grew. It was soon working with Atlantic Renewable Energy on a large wind farm in Iowa, several times the size of the two Pennsylvania projects combined. Located on a high ridge amid acres of corn, it demonstrated the skill that Zilkha Renewable was developing at mastering the industry's complexity. It installed ninety turbines, each atop a forty-two-foot-wide foundation of reinforced steel and concrete. The company purchased a historic barn and moved it six miles to serve as an office for the site manager and a warehouse for parts and tools.

Forty-nine property owners signed leases for the wind farm. Each received annual payments. The amount they received, one farmer said in a video produced years later by the company, was much more than he got from planting corn or soybeans. "We need something else besides corn," said another farmer. "We seem to have plenty of corn around the country." The project cost $80 million.

After Iowa, the company pursued bigger projects. The monthly and weekly calls for more cash got larger and larger. Zilkha Renewable partnered with PPM Energy to develop a large wind farm called Maple Ridge in upstate New York, and started another in Illinois.

As the pipeline grew, so did the calls on Michael and Selim Zilkha's checkbooks. One day in early 2004, Michael Skelly sat down and figured out what was coming due in the next twelve months. The bill had grown to $500 million. Michael Zilkha realized that the company had become too successful—and too expensive—to run as they had been.

"It was a joy to build it. An absolute joy to build," says Michael Zilkha. "As long as we could write the checks, we wrote the checks. Because we believed in it." But they reached a point where they could no longer write the checks. After some early stumbles, Zilkha Renewable, with Skelly and Winsor leading the way to find new places to build wind, mastered the complexities of the business. The pace of development picked up. So did the cost. Hippies were no longer running the wind business. Green pieties had been replaced by accountants' green eyeshades.

5

East 11th Street

S ince the days of Thomas Edison's Pearl Street power plant, power flowed one way: outward from the coal plants and hydroelectric dams. Customers generating their own power and sending their surpluses back onto the grid? Unthinkable, utilities contended. The voltage would collapse and the grid would black out. To keep the grid running, control room engineers should control the flow of power. This way of looking at the electrical world didn't leave room for wind turbines or solar panels.

But in the 1970s, an experiment showed there could be a new way of thinking about electricity and grids. Neither a national laboratory nor a forward-thinking utility ran the experiment. It was the work of a group of young men atop a tenement building in New York City's Lower East Side. The unwitting guinea pig was Con Edison. The utility never agreed to the experiment and didn't find out about it until it was well under way.

Con Ed grew out of Edison's original coal-burning plant at 255 Pearl Street. It was located on the northern edge of Manhattan's financial district, where today the hulking skyscrapers that house global financial firms begin to thin out. The Pearl Street coal plant has vanished without a trace or even a historical marker secured to a wall. A parking lot filled with Wall Street bankers' cars is all that remains.

Eventually the plants got bigger and more efficient, but the business model remained. Edison controlled the power generation and the wires. It was a closed operating system. Access to the grid was tightly policed.

How that system was cracked open is a largely untold story. It takes place atop a five-floor tenement building at 519 East 11th Street, two miles north of Pearl Street. The protagonists are a bunch of architecture students and community activists who erected a wind turbine amid the tar and masonry on the roof. They wanted to do battle with the mighty Con Ed. They ended up challenging nearly a century of legal precedents and business deals, and ushered in an era of renewable energy.

Ted Finch grew up on the South Shore of Long Island, where his father taught him how to captain a wooden-hulled gaff-rigged sailboat. He later took up wind surfing. "I had a real love of all that," he said. "The whole idea of lift and drag, vectors and force."

He spent two years at Princeton University before transferring to Hampshire College in Massachusetts, where he began to study wind and machines that could harness it. He talked his way into a graduate-level engineering lab at the nearby University of Massachusetts at Amherst taught by William E. Heronemus, a former U.S. Navy captain-turned-wind-zealot. In 1974, Heronemus proposed building exactly 13,695 wind turbines off the coast of Massachusetts, mounted on floating platforms like those used to drill for oil. It would have cost an astronomical $22 billion and was never seriously considered. But it captured the imagination of young idealists. For a school project, Finch built a twenty-five-foot wind turbine and erected it outside the physical plant building on Hampshire's campus. To his delight, it generated power, which he used to fill up a large battery. But he worried it might topple over if he wasn't there to maintain it, so he dismantled it before leaving campus in early 1976.

He moved to New York City that spring. He loved how gusts of wind whipped around buildings, and he dreamed of one day suspending wind turbines across the city's avenues to capture this raw energy. He thought another good place would be the flat roofs of the Twin Towers, then the newest additions to the city's skyline. Their distinctive vertical steel ribs helped brace the buildings from winds coming off the harbor.

It wasn't long before Finch heard that a group was rehabilitating a tenement on 11th Street. "I didn't know what I was going to do in New York City and it sounded interesting," he said. He walked over one day and went inside.

The building had housed several families until a series of fires left it uninhabitable in the early 1970s. The tenants moved out and the landlord stopped paying taxes around 1974. The oil price shock in 1973 drove up the cost of heating oil and depressed the economy. Landlords in the poorer neighborhoods couldn't raise rents enough to cover the rising cost of heating the buildings. Some resorted to arson, hoping to collect insurance when the economics of owning a tenement building seemed bleakest. The city was losing its housing stock at an alarming rate

Some of the city's activists wanted to reverse this trend one building at a time. One of their earliest efforts was at 519 East 11th Street. They took out a $177,000 loan from the city to renovate the building. They would put their sweat equity into fixing it up. When finished, they got to own the building and live there. The future tenants stripped the building down to its joists and beams. It was slow and dangerous work. Fire damage left many floors in precarious states. "Workers constantly had to avoid holes and weak spots in floors, and prevent piles of rubbish from accumulating and putting excessive loads on one area of floor," said Alva "Chip" Tabor III, a Yale architecture student who worked on the rehabilitation of the building and wrote about it for his master's thesis.

Despite its neglect, the building was beautiful and richly ornamented. Riveted pilasters flanked the third- and fourth-floor windows; underneath these windows were grotesque terra-cotta faces. Shafts allowed sunlight and fresh air into all the apartments. This attention to detail and living conditions set the building apart from many of the era's other slapdash tenements.

This particular block of East 11th Street, between Avenues A and B, was then known as "Strippers Row." Thieves took apart cars and resold the parts. On a productive night, twenty cars might be stripped. "It looked like Dresden or something," remembered Tabor. Drugs were readily available. "You couldn't walk two steps without tripping over a junkie or a pimp," an unnamed local told a journalist in the 1970s.

That was the neighborhood that Ted Finch navigated to get to the tenement. He had a wide mustache, longish hair, wire-rim glasses, and "a good Waspy jaw," said Travis Price, a solar energy advocate from New Mexico who was helping to gut and restore the building. Price and others formed what they called the Energy Task Force, as part of the rehabilitation, to devise ways to use less heating oil and therefore make it more affordable to live there. The group added insulation and installed a solar hot-water heater on the roof.

People who had been working on the house remember Finch as a gear-head. Nearly from the moment he walked in the door, he pushed a specific idea. Why not put a wind turbine on the roof? "He came in like a fireball," remembered Price.

While other members of the Energy Task Force could become passionate talking about community organizing and cutting energy costs for the urban poor, Finch focused on the technical issues surrounding connecting a wind turbine, including making sure the electrical current matched the power grid.

By the summer, the Energy Task Force decided to give it a try. Con Ed's electricity rates were close to the highest in the country. A wind turbine was "a safe hedge against even more drastic energy price escalations which threaten to wreak havoc in low-income communities," explained Finch and several others in a 1977 report on the project. Generating enough power to keep the lights in the hallways on meant purchasing fewer expensive kilowatt hours from the utility.

Finch had a second motive. He wanted to turn the grid into a two-way street. It had always run in one direction from power plants to power consumers. He wanted to add a second lane heading in the opposite direction. The effort was never about detaching from the grid and creating some renewable utopia. It was about forcing the utility companies to accept that renewables could be a contributor. "We were trying to fix the industrialized energy world, not roll it back," said Price.

Finch called Con Ed to ask about connecting his generator to the grid and got a long runaround. It was dangerous, they warned. It was illegal. It would destabilize the entire grid. It could fry the utilities' computerized control equipment. And in a Kafkaesque twist, Con Ed told him that no one knew what paperwork was required. Without the proper paperwork, it couldn't be done. "We never could find anyone at Con Ed who was the right one to talk to," Finch recalled.

Finch pressed on. Everything he had learned in the classroom and by building a turbine at college led him to the conclusion that installing a turbine wouldn't cause any grid problems. He figured the grid was so massive and diffuse, it would be possible to connect a turbine to it without anyone really knowing. The Energy Task Force purchased an old, refurbished turbine with a $4,000 federal grant. It was capable of a top output of 2 kilowatts. By comparison, Con Ed could hit a peak generation of 8,609,000 kilowatts, from natural gas, coal, and nuclear power plants.

Finch purchased a turbine manufactured by the Jacobs Wind Electric Co. In the 1920s, the Minnesota-based firm had applied propeller technology developed for airplanes in World War I to wind machines. The Jacobs turbine was an unsung mechanical wonder of the American century. These turbines provided the first glimmer of power to many farmhouses long before transmission wires and rural electric cooperatives. The company sold an estimated thirty thousand turbines across the Plains states from 1927 to 1957.

Jacobs turbines were hardy. In a 1973 interview, Marcellus Jacobs was asked about his machines' reputation for never breaking down. "I'm kind of a freak, see. I want things to work forever. I built my plants to last a lifetime," he said. But they couldn't survive the spread of the power grid into rural America. Jacobs testified before Congress in 1974 that the federal government's Rural Electrification Administration had doomed his machines. When wires arrived, farmers no longer wanted or needed their Jacobs turbines. Farmers retired thousands of them from use. Decades after most of these turbines were disconnected, a couple of entrepreneurs from Vermont hitchhiked through Colorado, Minnesota, and North Dakota, buying up used turbines from farmers, then refurbishing and selling them. One ended up headed to New York City, packed in a large crate.

Sometime in the fall of 1976, probably around October, Finch and his compatriots hoisted it up to the roof with a gas-powered winch. They planned to put it atop a thirty-seven-foot steel tower, which they also raised up in pieces. With a sixty-foot roof, the turbine would be one hundred feet above the ground.

Finch knew it wouldn't be enough. Over the summer, he had set up a telescoping tower with an anemometer to measure wind speed. After a few weeks, the equipment reported back that average wind speed was around 10 miles per hour. "It really didn't justify putting a turbine up," Finch told me nearly four decades later. "You really needed 13 or 14 miles per hour. I knew the economics wouldn't be there."

But the Energy Task Force didn't care. "The precedent wasn't that we put up a little two-kilowatt wind generator," Finch said. "The precedent was being able to make your own power." When the wind was strong and the turbine made more power than could be consumed by the building, Finch wanted to flow it back to the power grid and feed it to Con Ed. The twenty-three-year-old Finch wanted to use a $4,000 wind turbine to do battle with a company that had an electrical plant investment worth $6.7 billion, and incalculable political clout.

Led by Finch, a group from the tenement erected a metal tower made of heavy galvanized steel tubes. They assembled it into two pieces and, like an urban barn-raising, put up a thirty-foot section of the tower. A black-and-white photograph shows about a dozen men pulling on ropes attached to the top. Another few are busy around the base, trying to get the tower connected to long bolts coming out of a recently poured concrete base. It appears to be a beautiful day with only a handful of wispy clouds in the sky. The Twin Towers are visible in the distance. The group attached a second, smaller piece of the tower and hoisted up the bladed generator by pulley. Finally, they attached the blades and tail vane.

Finch wired it into a large red inverter box, which turned the direct current into alternating current. Despite Con Ed's warnings that they could lose control of a power grid that had begun in 1882 and grown to serve eight million customers, Finch hooked up the turbine to allow excess power to flow back to the grid.

Finch and several other members of the Energy Task Force gathered downstairs in the basement to see something that, quite possibly, no one had ever witnessed before. "We thought it was cool to watch the meter spinning backward," he said.

———————

The grid didn't collapse. There was no blackout. (When a large blackout plunged New York City into darkness in July 1977, the tenement was one of the few points of light visible in downtown Manhattan from passing airplanes.) Nobody from Con Ed seemed to have noticed the tiny new generator on its system. The turbine was intended to provoke a reaction, and by this measure it was a failure. So the Energy Task Force issued a press release on November 12, 1976.

The next day, the *New York Daily News* ran a photograph of Ted Finch climbing the tower with the hulky Con Ed power plant on the

East River behind him. The *MacNeil/Lehrer Report* aired a national piece on the wind turbine. *The New York Times* ran the headline: "11th Street Tenants Tilt with Windmill and Con Edison." The articles imply that the system had yet to be connected, but Finch says this was a bit of subterfuge. It had already been connected. And as the long-haired energy activists on East 11th Street became media darlings, Con Ed didn't know how to react.

At first, Con Ed threatened to force Finch and the rest of the group to disconnect the wind turbine, but they had powerful friends. Congressman Ed Koch, who would go on to be mayor for twelve years, had toured the tenement earlier in 1976 and praised the insulation and solar hot-water collectors the group had installed as a "promising source of energy." Congressman Richard Ottinger had also toured the building, heaping praise on the effort to generate energy.

The tenement and its urban wind turbine became a bit of a cause célèbre. Ramsey Clark, U.S. attorney general under President Lyndon Johnson, offered to be their lawyer. But the Energy Task Force chose a different lawyer named John B. O'Sullivan to help them battle Con Ed.

O'Sullivan always looked out of place when he arrived for meetings on East 11th Street dressed in a suit. He came from a wealthy and politically connected family. In 1969, he hosted a gathering of antiwar, politically active young men at his parents' home on Martha's Vineyard. One of the attendees was a young Rhodes Scholar named Bill Clinton who spent the weekend fretting about his draft status. Mocking the pretensions of young men, another attendee referred to the group as "The Executive Committee of the Future."

O'Sullivan's first job as the Energy Task Force's lawyer was to respond to a filing by Con Ed with the state utility regulatory commission in January 1977. The utility's strategy was to bend, but not break. It offered a proposal that recognized the ability of people to generate their own electricity and even sell some back to the grid. This was a stunning move, and it reversed decades of utilities arguing that they should have complete control of power generation and distribution. But the proposal also made it difficult, and likely impossible, for anyone to replicate what had been done on East 11th Street.

Con Ed proposed that any new wind turbine owner must provide an "oscillographic print" to demonstrate that the current and voltage were within normal ranges. And they needed to provide information on "percent wave-shape distortion." They also proposed a liability clause that would have left the wind turbine owner responsible for any and all mishaps. The final impediment was an exorbitant minimal electricity charge. O'Sullivan deflected Con Ed's attacks with a simple argument. Responsible groups should have the ability to generate their own power and sell what they didn't use back to the grid.

In May 1977, the state public utility commission issued its ruling and sided with Finch and the Energy Task Force. The wind turbine was legal and Con Ed was required to take the power. The liability clause was reduced, as was the requirement for oscillographic prints. The minimum charge was slashed. The East 11th Street tenement association would pay a $1 a month charge for a meter to track the electricity that flowed back to the grid. And they would be paid 2.3 cents per kilowatt hour. At the time, Finch estimated they had been sending about 20 kilowatt hours back per month, earning them 46 cents. But their rooftop electricity let them use less utility power and avoid paying $17 a month to Con Ed.

When they heard the news, Finch and others climbed up on the roof and celebrated. Roberto Nazario, a local activist who had led the effort to restore the tenement, penned a free verse poem to commemorate the triumph.

There are three legal, recognized power companies in New York City:

Con Edison
Brooklyn Union Gas
and the people of 519 East 11th Street.

The utility couldn't help taking a little dig at the renewable pioneers. "If the wind stops for a minute," a company spokesman said, "we'll be there."

Between the relatively low wind speeds and the meager returns, the rooftop turbine generated a lackluster return. "It proved that doing so in an urban environment was not a good investment. It was probably much smarter to do it on a ridgeline in Montana," says David Norris, a student who worked with the Energy Task Force, and went on to advise architects on using natural light effectively in their buildings.

It was a narrow victory, involving a vanishingly small wind turbine in a giant pool of power, and the decision only affected one utility. But it established the principle. A person or tenant cooperative or company could generate electricity and sell it to their local utility. The grip of the utility was loosened.

At the time, Congressman Richard Ottinger, a longtime mentor and friend of O'Sullivan, was helping push a bill through Congress called the Public Utility Regulatory Policies Act, or PURPA. Ottinger's law encouraged small-scale power producers—particularly renewable energy companies—to enter the market and required utilities to purchase their power at a fair price.

It was a legislative response to the energy crisis. The common conception at the time was that natural gas supplies were running out. Nuclear power plant construction was running over-budget. Burning oil generated a lot of electricity, and a boycott led by Saudi Arabia and other members of the Organization of Petroleum Exporting Countries (OPEC) was driving up the price of oil and using control of supplies as political muscle. Lawmakers wanted to open up the power market to new domestic power producers.

In November 1978, President Jimmy Carter signed PURPA into law. Once unassailable, power companies now had to make room for newcomers. The utilities still provided the overwhelming bulk of electricity, but others were allowed into the industry. Paper mills and chemical plants, which generated lots of heat from their operations, could use this excess to generate electricity. Companies could harness the wind and sun to make power. These new generators could sell their power to the utilities, which were required to buy it and distribute it.

It was a scratch on the monolith. Over time, that scratch would widen and be eroded by subsequent changes until the integrated utility, a single company that handled all aspects of electrical generation and distribution, would be in retreat. "I don't think anybody had any idea that this would be the start of the total restructuring of the industry," said Bob Shapiro, a former federal lawyer involved in implementing the new law.

While Congress mandated changes, the law was silent on a crucial detail. How much would these new power providers get paid for their electricity? The job of devising rules to govern the new power transactions was left to the year-old Federal Energy Regulatory Commission. Robert Nordhaus, the first general counsel of FERC, wanted help writing the rules. He had heard about John O'Sullivan and his battle with Con Ed. He hired him before Congress passed PURPA.

For a year, O'Sullivan worked on the rules. He was aware that utilities, if unchecked, were unassailable. They could make life difficult for the new power providers, just as Con Ed had tried to do on East 11th Street. Before the passage of the new law, utilities could simply refuse to purchase power from anyone else, or offer to pay a pittance. If the rules weren't robust, utilities could make it hard for the new power generators to connect to the grid. And they could refuse to pay a reasonable price for the power, financially hobbling the new industry.

O'Sullivan worked to make sure PURPA wasn't an empty victory. He pushed for the maximum reimbursement rates. While not delivering any savings to consumers in the short term, these rules would create new sources of power that would drive down prices over time. They would create competition. "The nation as a whole will benefit from the decreased reliance of scarce fossil fuels, such as oil and gas, and the more efficient use of energy," O'Sullivan wrote.

When they read FERC's rules, the utilities went ballistic. American Electric Power, a giant Midwestern power company, and Con Ed sued the government. A three-judge panel of the U.S. Court of Appeals heard the case. They gutted O'Sullivan's rules in a decision handed down in January 1982. FERC appealed the decision to the Supreme Court.

On May 1983, nearly seven years after Ted Finch hoisted up the wind turbine, the Supreme Court handed down a unanimous decision. They reversed the lower court and handed a resounding defeat to the utilities. It was a complete validation of O'Sullivan's rules. The court, in its decision, copied word for word the rationale Sullivan had laid out a couple years earlier. "The nation as a whole will benefit from the decreased reliance of scarce fossil fuels, such as oil and gas, and the more efficient use of energy," the court wrote. By this time, O'Sullivan had left government employment and begun a long career at the law firm Chadbourne & Parke. He died in 2010.

Finch's turbine, PURPA, O'Sullivan's rules, and the Supreme Court decision helped create a new industry of independent power producers. "It produced quite a boom in new types of generation," says David Spence, a professor at the University of Texas at Austin who teaches energy, regulation, and law. Entrepreneurs entered the market experimenting with ways to generate power, first burning gas and later building large-scale wind and solar farms. States also began to experiment with new ways of regulating power markets. In 1983, Iowa introduced the nation's first renewable portfolio standard law, requiring utilities to generate either a certain amount of renewable power or buy it from an independent company. Eventually thirty states passed similar laws.

In 1992, Congress enacted a sweeping energy law. If PURPA had chipped away at the utility companies' monopoly, the Energy Policy Act of 1992 took a sledgehammer to the old model by opening the wholesale electricity market to competition. FERC followed up with rules to implement the law and opened up the grid. Ted Finch's

2-kilowatt wind turbine had started a momentous change. The law now allowed independent power companies to generate power and sell it on the bulk power markets. Several years later, Skelly and Zilkha Renewable would take advantage of these new rules to put more wind power on the grid.

––––––––––

The 11th Street wind turbine is long gone. Ted Finch says he took the turbine down around 1980. No one was keeping an eye on it, and he worried it would break and a wayward blade could fly through someone's window or land on the street. All that's left are a few eyeholes and steel rods threaded into the walls that once anchored the turbine.

Directly north of 519 East 11th Street there are four mounted arrays of photovoltaic panels, maybe capable of generating 5 kilowatts—two and a half times more electricity than the size of Finch's wind turbine. That solar installation is a direct descendant of Ted Finch's wind turbine.

6

The Battle of Blue Canyon

J ack Kline returned to his hotel room dog-tired from a long day. He had flown into Oklahoma City, rented a car, and headed toward the southwest corner of the state.

He picked up the phone and dialed Michael Skelly. Kline chose his words carefully. He knew the wind developer was busy with multiple projects and he didn't want this opportunity to get lost.

"You have to go for this," the wind prospector said. "This is gold. Don't blow it."

A few weeks earlier, he had noticed something promising while scanning a large topographic map of Oklahoma. Some contours caught his eye. They looked interesting, he thought, and suggested hills there might concentrate the wind. He wanted to see it in person to be sure.

After a couple of hours driving across Oklahoma, he arrived at the spot on the map. "I spied this ridge and my eyes popped out of my head," he said. It was possibly the best location for a wind farm he had ever seen.

Kline had first met Skelly in Costa Rica. Skelly had hired him to study how wind whipped across Lake Arenal and determine the optimal turbine placement. Common sense would suggest the best place for a wind farm is where the wind blows. But nothing is ever that simple. And while the wind industry was still in its early years, before maps built from a thousand data points simplified the search, smart developers needed help telling the difference between the merely windy and a great wind location. It could mean the difference between striking it rich and running out of cash.

Kline started in the business back in the early 1980s when little was known about wind. By 2001, when he began scouting, he had a weathered face and crow's-feet around his eyes, the look of a man who had spent many of his days outdoors. A meteorologist by training, he wasn't particularly interested in theoretical disputes. He liked to measure the wind. He wanted to understand how it moved across terrain and through a wind farm. He studied the magnitude and frequency of gusts.

He schooled a nascent wind industry on the turbulence created by turbines. It was once commonplace to build several rows of turbines packed together facing the wind. Kline helped show that turbines created a wake, like a motorboat crossing a lake. Passing through a turbine, the wind eddied and rumpled. The first row of turbines performed as expected. Subsequent rows would produce less power and the turbines would fall apart more quickly. Kline helped map wake turbulence, leading to new and improved turbine spacing.

But writing papers ("Field Comparison of Maximum Cup, Climatronics and Met One Anemometers") didn't pay the bills. In the early days of Zilkha Renewable, Skelly supplied Kline with steady work because Kline had a knack for wind prospecting. Even into the 2000s, some people still looked for wind by scouring maps for a "Windy Hill Road" or something similar. Kline, with a master's degree in atmospheric sciences from Georgia Tech, brought a more scientifically savvy perspective.

Skelly sent him to Michigan, Wisconsin, and Illinois. Kline would drive around looking for mesas or hills. These features amplified the wind, making it stronger and more consistent. In southwestern Oklahoma, Kline thought he had found something special.

What Kline found was an area north of Lawton known as the Slick Hills. The rock-strewn ridges rise 400 to 1,000 feet above a flat expanse of grassland. Locals knew that the hilltops were windy. Randy Gilliland, whose grandfather bought land in the area and passed it down to him, said the Slick Hills always seemed to be wind-whipped. "It is a really strange thing. We could be down at the bottom of the Slick Hills, fishing or just hanging out in the cabin, and there is no wind at all. And then you get in a four-wheeler and go up on top of the Slick Hills and you couldn't hold on to your hat," he said.

Zilkha Renewable was growing in size and ambition. Kline's message was all the encouragement Skelly needed. The company's Pennsylvania wind farms were each less than 20 megawatts. The one in Iowa was four times that size. The Slick Hills were large enough that a wind farm built there would be measured in the hundreds of megawatts. It was an exciting prospect.

Skelly dispatched Zilkha employees to the Slick Hills to generate interest among the landowners. They soon learned they weren't the only wind company knocking on ranch doors, and Rick Walker, the chief wind developer for a Dallas power company, beat Zilkha's representatives to the area. "The wind was screaming up there," said Walker. The Dallas company had done some advanced planning and concluded that wind was becoming cost effective. "We wanted to find the best locations" and lease them, he said.

At the time, there weren't any wind farms in Oklahoma. Now, not one, but two companies wanted to lease land from Gilliland and his neighbors. It all seemed "like a dream," he said. The Slick Hills were a good place to stumble on diamondback rattlesnakes, some as long as six feet and as thick as marine rope. While the surrounding flatlands were suitable for grazing cattle, no one could figure out any way to generate income from the bumps on the prairie. Trees wouldn't even grow there, leaving the hills with the odd nude appearance that led to their name. Years later, when a wind farm had been erected, Gilliland got a note from a fellow landowner. "Whoever dreamed we would be able to achieve this kind of income off our pile of rocks," it read.

At the time, Gilliland worried the turbines would spook the cattle and scare off the wild turkeys that he and his neighbors hunted in the fall. Gilliland called several out-of-state landowners who had wind farms on their properties. What he heard comforted him. The cattle weren't fazed. The hunting was still good. The giant swooshing blades didn't bother the animals.

Local landowners gathered at a run-down community center—located at the crossing of two roads with nothing else nearby—to discuss the idea and consider the competing proposals. Zilkha Renewable provided a steak dinner one night while making its pitch. It was the out-of-state company with the funny name. It wanted every advantage it could muster.

Skelly dispatched Wayne Walker to be his emissary in Oklahoma. Walker, no relation to Rick Walker, had worked in the semiconductor industry, selling chips to companies that made cell phone towers. He felt unfulfilled by the work and went back to school for a degree in environmental sciences. After graduating, Skelly recruited him to Zilkha Renewable. He found a company filled with people with business backgrounds who wanted to do something for the environment. People like him.

Walker loved the work. "It was exciting. We are on the tip of the spear of change—and everywhere we went there was resistance to change. People kept telling us we weren't real," he said. He met with state legislators, electricity cooperatives, and utilities to try and convince them to take wind power seriously. He spent every weekday in Oklahoma, figuring out how to win over the landowners. He soon sized up the situation: there were numerous landowners who held a piece of the Slick Hills, but only one that mattered. Right in the middle of the hills, the Kimbell Ranch covered about fifteen thousand acres. Without the Kimbells, there would be no large wind farm. Get them on board and other landowners would follow, Walker figured: "If you got the Kimbells, you got the mountain."

But the Kimbells, a family of Texas oilmen, were skeptical of renewable energy. Three generations of Kimbells had owned and run the family oil company, Burk Royalty. They drilled wells all over Texas and Oklahoma. G. T. Kimbell had chartered the company in 1935, listing a then impressive $100,000 in capital stock. Mostly an oil and gas driller, the company had also owned several small refineries.

Wayne Walker was relentless. Stan Kimbell, the founder's grandson, recalled times when he was out hunting and Walker would drive up, park his car, hop the fence, and amble over to visit. Kimbell, who speaks with a leisurely Texas drawl, worried about the ranch, which had been in the family for fifty years. In 2004, an association of state fish and wildlife agencies honored him as the "Landowner of the Year." His progressive stewardship included using an advanced rotating grazing system for the cattle to make sure bird habitat was preserved. The ranch had the largest private elk herd in the state. It is "legacy property," he said.

The Kimbells were interested, but cautious. Stan Kimbell drove a few hours to West Texas and walked around some new wind farms there. The size of the turbines stunned him. He also took an online course on wind energy. "I began to see the reality, that this really is a commercial possibility. It is practical and it is proven," he said. "The size was something that was extremely new to me. The top of the blade was 320 feet in the air. It was a major, major piece of equipment."

Skelly visited him and his father, David, several times at Burk headquarters in Wichita Falls, Texas, an hour south of the Slick Hills. Kimbell found Skelly impressive, and he liked that the Zilkhas had been in the oil business. "It gave us a level of comfort with them," Stan Kimbell said. He liked that he could pick up the phone and talk to the owners and top people at Zilkha Renewable. "We just felt like it was maybe a little less corporate" than the competing proposal from the wind unit of American Electric Power. "You know, we felt it was a better fit for us."

Stan Kimbell had one condition. At Burk Royalty, he was used to being the oil well operator. He signed the royalty checks and sent them off to landowners. The Kimbells were willing to host the wind farm if they were part owner, he told Skelly.

Skelly took the message to Houston and sat down with Joe Romano, the Zilkha Renewable chief executive. Zilkha Renewable was a wind development company. As chief development officer, Skelly was the company's visionary, strategist, and cheerleader, but Romano had experience running a multimillion-dollar company. Skelly always wanted to go faster, and Romano's job was to figure out when to slow him down and when to let him run.

Without the Kimbells, there would be no wind farm at the Slick Hills, Skelly explained. Romano suggested a joint venture; bring in the Kimbells as part owners, as they preferred. This was the best way quickly and decisively to lock up the land, so construction could begin. The Zilkhas signed off on the idea.

Skelly offered, and the Kimbells accepted, a 25 percent stake in the wind farm, which would be called Blue Canyon. Together, they bought out another family ranch owned by two feuding sisters who wanted nothing to do with each other or the land.

Wayne Walker moved into the ranch house and lived there during the workweek for six months as he continued signing up landowners. With the Kimbells on board, his job became easier. He had soon leased up enough acres to build a large wind farm, one of the largest in the United States.

The Kimbells grew nervous at the pace of spending. They scrutinized every expense. The bills were large and some spending seemed to them unnecessary. "There was no assurance there was a light at the end of the tunnel and the costs were much larger than we had budgeted," said Stan Kimbell. At one point, Zilkha Renewable placed a multimillion-dollar wind turbine order before getting the Kimbells to sign off. Stan Kimbell told Skelly and Romano this wasn't how his family operated. Their hastily arranged marriage was making him increasingly uncomfortable.

Once again, the Zilkha checkbook came to the rescue. Skelly renegotiated the terms. The Kimbells accepted a reduced 20 percent ownership stake; in exchange, Zilkha Renewable agreed to pay all of the expenses.

Getting the land leased was the easy part. Oklahoma required that the output from Blue Canyon had to be sold to local utilities. There was no market that Zilkha Renewable could bid into to sell the electricity from the hundreds of planned turbines. There were commodity markets for crude oil, gold, lean hogs, soybeans, even frozen concentrated orange juice, where buyers and sellers could find each other and let supply and demand determine the price. But there were no markets for power in Oklahoma, or in most of the rest of the country.

Only a few local utilities in that part of Oklahoma could buy the electricity generated by the Blue Canyon wind farm. Zilkha Renewable negotiated to sell a slice of the output to a local electric cooperative. The other possible buyer was Public Service Company of Oklahoma, a subsidiary of American Electric Power. AEP had just merged with Central and South West, adding Public Service to its stable of regional utilities. At the time, it was the largest power company in the United States, a giant corporate octopus that controlled power generation and transmission from the hollows of West Virginia to the plains of West Texas.

AEP wasn't particularly interested in buying power that someone else had generated. It owned hundreds of coal and gas plants around the country, burning enormous amounts of fossil fuels to spin magnets around and generate electricity. It even owned a small number of wind farms. It had millions of captive customers and decades of experience working the state regulators to keep its profits secure. AEP bristled at what Blue Canyon stood for: an erosion of the giant power monopoly that AEP had spent billions of dollars to build.

"AEP hated independent power producers," said Denise Bode, who at the time was an elected member of the Oklahoma Corporation Commission, which regulated utilities. "AEP hated to be forced to take any kind of independent power production."

When an independent company proposed building a natural gas plant in nearby Lawton, AEP dispatched several lawyers and economists from its headquarters in Ohio. They descended on Oklahoma City, providing hours of sworn testimony before the Oklahoma Corporation Commission to explain why this arrangement was a bad idea. "They fought tooth and toenail," said Bode.

It didn't matter if the electricity came from gas turbines or wind turbines. AEP didn't like an independent company generating megawatts in Public Service Company of Oklahoma's backyard.

———————————

Frederick William Insull, a British-born businessman with a surname that evoked twentieth-century electricity royalty, founded Public Service Company of Oklahoma in 1913. Fred, as he was called, was a nephew of Samuel Insull, the Chicago business tycoon and Thomas Edison's former private secretary. The uncle built an empire upon a simple belief: electrical power should be affordable for the masses but owned by a small clique. By the end of the 1930s, his power empire was cut down, first by the stock market crash and then by the trust-busting Public Utility Holding Company Act.

Neither of these setbacks dulled Fred Insull's ambitions in Oklahoma. PSO, as the company was sometimes known, consolidated power companies throughout the eastern half of the state. In 1944, it added Southwestern Light and Power, giving it the same corner of the state as the Slick Hills. The investment soon paid off. The region boomed during and after World War II, in part because of wartime activity around Fort Sill. The census counted 18,055 people in Lawton in

1940. A decade later, there were 34,757 and 61,697 a decade after that. To keep up with growing electricity demand, PSO built a natural gas power plant in nearby Anadarko, Oklahoma. The first unit opened in 1950, a second in 1952, and a third unit in 1967.

These units were glorified kettles, burning gas to make steam and using the steam to generate electricity. By 2002, they were antiques that required costly maintenance. Whenever demand was high and PSO ran these units, the combination of expensive fuel and old machines made for expensive electricity. The company passed these higher costs on to its customers. Even if the power was expensive, PSO still made a profit.

In November 2001, Skelly tried to initiate talks with AEP about buying the output from the as-yet-unbuilt Blue Canyon. He kept getting passed around from PSO's headquarters in Tulsa to AEP executives in Columbus, Ohio, and then back again to Tulsa. Skelly understood the message the company was delivering. It wasn't interested in making things easy for his wind farm.

When Zilkha Renewable filed a request to build an interconnection between the wind farm and the main grid, AEP claimed it would cost $19.9 million, even though a similar nearby interconnection only cost $5.3 million. AEP argued that the interconnection required upgrading a different power line near Tulsa, on the other side of the state. Zilkha countered that AEP was padding the costs and trying to get the wind company to pay for unrelated system upgrades.

Knowing there would be a fight ahead, Skelly started working on the state regulators. He flew to Oklahoma City to meet with Denise Bode. He figured she would be a hard sell. She was whip-smart, politically savvy, and from a petroleum family. Her father had been an executive at Phillips Petroleum. Before she was elected to the commission, she had lived in Washington, D.C., where she ran the Independent Petroleum Association of America, a trade and lobbying group, for seven years. Her office decor didn't set him at ease. Her window faced the state capitol and a working Phillips Petroleum pump jack in a flower bed across the street. There was an empty bottle of Jim Beam bourbon with a taped-on picture of J. R. Ewing. There was also a two-foot replica of the "Oilfield Warrior" statue. The original statue was erected in Nottingham, England, and commemorated the little-known story of Oklahoma roughnecks sent to England during World War II. To help the war effort, they drilled dozens of wells in the country's only oilfield, boosting production tenfold.

Skelly talked about his vision for wind farms in Oklahoma and how it could help the state. "He was very articulate and very knowledgeable," she recalled. He explained how costs were coming down. He said it would be a good deal for consumers, shielding them from rising and falling natural gas and coal prices. "He came in with a vision of what it was going to be," she said. She was intrigued and would soon become a convert.

Two years later, when the state's other big regulated power company, Oklahoma Gas & Electric, created a program to allow its customers to pay a little extra in exchange for supporting wind farms, Bode was the first Oklahoman to sign up. She put a sticker on her window at home. A petroleum industry chief executive, at her house for a fundraiser, saw it, and sneered. "I hate wind," he told her. Others loved it.

It turned out to be a good deal. At first, Bode and others who signed up paid more for their electricity than neighbors who stayed in the traditional plan. But Bode didn't have to pay a fuel-adjustment charge that rose and fell with the price of natural gas and coal. By 2005, the fuel charge had risen, but the wind price hadn't, and Bode paid less on her power bill than her neighbors buying electricity from fossil fuels and nuclear plants.

This was almost exactly what Skelly had predicted in her office in the early 2000s. By the end of the decade, Bode moved back to Washington, D.C., to run another industry trade and lobbying group. This time, it was the American Wind Energy Association.

Skelly's attempts to negotiate a deal to sell Blue Canyon's electricity to AEP were going nowhere. If AEP wouldn't sign a deal to purchase the output, Zilkha Renewable could bring a regulatory action before the Oklahoma Corporation Commission and let them sort it out. Michael Zilkha didn't think this was the right approach. In their oil business, the Zilkhas had a way of making friends out of enemies. Michael Zilkha hated the idea of starting a business relationship with a customer—they were trying to sell power to AEP, after all—by bringing out lawyers and forcing everyone to give depositions. Skelly agreed with him. Wayne Walker wanted to fight. At least once a week, he pressed Skelly and Zilkha to treat AEP as an adversary. "We can beat these guys," he said.

Skelly started looking at how PSO operated in Oklahoma, studying what power plants the company deployed. He started to get angry. And the more he looked, the angrier he became. "They were running a lot of inefficient gas plants during a time of high gas prices," Skelly recalled. "Our wind would save their ratepayers money."

Some calculations bear out his insight. The heat rate—literally the amount of natural gas that a plant needs to burn to generate a kilowatt hour of electricity—was striking. In sworn testimony, a PSO official in 2002 said that Unit 3, the newest and most modern, had a heat rate of 11,500 British thermal units; the federal government reported a heat rate of 11,705 BTUs. Units 1 and 2 were older and less efficient. Based on gas prices in 2002, this meant PSO could generate a kilowatt of power for 4.1 cents; a year later, higher gas prices would raise that to 6.5 cents. Skelly's first offer was to sell wind output for 4.3 cents, but he knew they could go lower, close to 3 cents, and still make a profit.

"We saw the heat rates on these old, stinky gas units, and I realized we were going to save Oklahomans a lot of money," he said. But those savings would only materialize if AEP agreed to buy the power. If they wouldn't agree, they could be so compelled. Zilkha Renewable could file a case under the the Public Utility Regulatory Policies Act, PURPA. Bringing a PURPA case would result in a drawn-out legal and bureaucratic mess that only a utility lawyer who charged by the hour could love. But it seemed to be the only way to force the company to sit down at the negotiating table. And without AEP as a buyer, there would be no turbines atop the Slick Hills.

The only way to get the wind farm built, Skelly decided, was to fight AEP. He adopted Wayne Walker's position that Zilkha Renewable should file a case to force AEP to take the megawatts generated from the wind that slid across the Slick Hills. Bringing a case was the only way to get AEP to sit down and negotiate.

On July 8, 2003, Zilkha Renewable wrote a formal letter to AEP requesting they initiate formal negotiations for the sale of power. It was the required first step in the legal case to compel AEP to buy their power. More than two years had passed since Jack Kline had called Skelly about the Slick Hills.

In October, AEP promised to provide a draft of a legal agreement to purchase power. They didn't. By the end of the month, Zilkha Renewable had filed an application with the Oklahoma Corporation Commission to compel the utility to negotiate a deal. Zilkha Renewable, the application argued, is a capable wind developer that had spent more than $200 million building wind farms. It had nearly thirteen thousand acres leased. "The project is viable, well progressed, well invested, well known and well supported," the company argued. It had a construction contractor and a deal to purchase turbines. It had completed an environmental assessment and notified nearby Native American tribes to ensure that cultural sites wouldn't be disturbed. All that stood in the way was AEP. A week and a half after the case was filed, AEP submitted a power purchase offer, about twenty months after Zilkha Renewable had first requested it.

The case was assigned to an administrative law judge, who ordered the sides to keep negotiating. Whether Zilkha Renewable won or lost depended on how the judge ruled on the issue of "avoided cost." If the wind farm cost was lower than the avoided cost of running the gas plants—thereby saving electricity purchasers money—then AEP would be compelled to purchase the power. But exactly how to define avoided cost? "There wasn't a lot of precedents, there wasn't a lot of understanding of what avoided cost meant," said Cheryl Vaught, the Oklahoma City attorney hired by Zilkha Renewable. "We were somewhat in a new frontier with this." Zilkha Renewable had a couple of lawyers on their case. AEP mustered a dozen or more.

Skelly's passion impressed Vaught. "He would never get down about anything," she said. "He was just the most enthusiastic human being you could ever meet, while at the same time being a businessperson. Which I just loved that about him. It is never all about us. It was about what is right and what is logical and what is good." By the spring of 2004, Skelly was getting antsy. "We are developers. We want to build stuff," he said.

One day that spring, the AEP lawyers and technical experts flew from Ohio to Houston on one of the AEP corporate jets for negotiations scheduled to begin at lunchtime. Talks were to be held in the Zilkha offices, in the conference room presided over by the oversized picture of a glowering Johnny Rotten.

That morning, Wayne Walker and the Michaels—Skelly and Zilkha—met to discuss strategy. Walker and Skelly had biked in to work together that morning. Michael Zilkha had also biked in. All three wore form-fitting, polyester-spandex bike shorts and shirts. The meeting wound down and the men prepared to take showers and change into suits for the negotiations.

Skelly had an idea. Why not insert a wild card into the talks, just to throw AEP off its game? Why not act a little bizarrely to send a message that they were a little irrational and would act irrationally until they were able to accomplish their goal of building a wind farm?

"Don't change. Don't shower," Skelly said to Michael Zilkha. "Go to the negotiations wearing your bright-pink biking clothes. They need to know that we are going the distance on this thing," he said.

Wayne Walker and Skelly took showers and changed into attire expected from a modern American corporate negotiation: slacks, suits, ties. Michael Zilkha stayed in his pink shorts.

AEP's lawyers and negotiators showed up on time and sat around the large conference table underneath Johnny Rotten's watchful and slightly crazed eyes. Cheryl Vaught was there, prepared to argue the fine points of PURPA law and power plant heat rates. Wayne Walker and Skelly said their hellos to everyone, making sure the sandwich platters and drinks were set up for their guests. They all grabbed food and sat at the table, ready for a few hours of court-required negotiations. The administrative law judge assigned to the case was expecting them to report back on the progress of talks within a couple of days.

Opposite the photograph was a wall of glass with a view down the hallway. A few minutes after everyone had assembled, Michael Zilkha strolled down the hallway in his pink ensemble. Few of the AEP officials had met Zilkha before and probably wondered who this middle-aged bike messenger was and why he was ambling awkwardly in his biking shoes into the meeting.

He plopped down into his chair and, according to Wayne Walker, swung his bike-shoed feet up onto the table. He began to lecture the stunned group about how bad coal was for the environment, and how gas burned in inefficient plants was only marginally better. "How can you live with yourselves?" Zilkha asked the AEP folks. He talked about how Blue Canyon would be cost competitive and how blocking it would be wrong.

After twenty minutes, his speech ended. Zilkha switched over to become a charming host. He thanked everyone for coming to Houston. He made sure they all had enough to eat. Then he left, walking back down the hallway. He didn't have to spell out his message. It was clear enough to the people in the room. They were dealing with a guy who was committed to his cause, and perhaps a little crazy. But he had money and wasn't planning to back down.

Negotiations between Zilkha Renewable and AEP continued, but it felt to Skelly and others that something had changed. Toward the end of 2004, the two sides began to make progress toward a deal. Zilkha Renewables lowered its price to about 3 cents a kilowatt hour for ten years and AEP agreed to buy the output. Skelly would get to build his wind farm, generate a lot of clean power, *and* reduce costs to Oklahoma customers.

On March 2, 2005, Public Service issued a press release that it would purchase up to 120.6 megawatts of power from Blue Canyon. "It allows us to provide our customers with the environmental benefits of clean, renewable energy while not raising PSO customers' rates," said company head Stuart Solomon.

A week later, PSO made another announcement. It would buy another 30.6 megawatts from Blue Canyon. Indeed, the size of the project had to expand by seventeen turbines, bringing the total to eighty-four. The Slick Hills wind farm would be one of the largest in the country.

There are hundreds of pages of testimony from Public Service of Oklahoma officials worrying that adding wind power from Blue Canyon—2 percent of the overall generation mix—would force PSO to buy expensive backup power. This didn't come to pass. Indeed, the utility liked wind so much it kept adding more. A decade after the Battle of Blue Canyon, PSO derived 22 percent of its power from wind and proudly boasted of a new solar installation on the campus of Tulsa University.

Like most power companies, Public Service of Oklahoma has to draw up long-term plans called "integrated resource plans." These days, its plan prominently features wind. By 2024, it expects 28 percent of its power to come from wind farms. And now that Skelly helped break down opposition and convince the company that renewable energy can work and work well, PSO expects to bring on more solar as well. It plans to add large fields of solar panels in the 2020s, getting 3 percent of its power from the sun. Two of the three inefficient units at the inefficient gas plant that angered Skelly are slated for retirement in 2021 and 2023.

The Southwest Power Pool, the multistate power coordinator that includes Oklahoma, Kansas, Nebraska, and parts of several other states, changed how it operates the power market in 2014. It now takes bids from power plants and dispatches the power it needs, beginning with the least expensive. The entire lawsuit with AEP would no longer be necessary, since Skelly could have built his wind farm and bid power into the grid.

In 2016, more than one of every 7 megawatts in the region came from wind turbines. In February 2017, for a few minutes, wind set a record when more than half of the power on the regional grid came from wind. It was the first time that wind cracked 50 percent on any of the country's large regional grids. A year later, in March 2018, wind briefly hit 62 percent.

7

A Cure for Insomnia

B y the summer of 2004, Michael Zilkha was having trouble sleeping. He had acquired a plucky company that built 20-megawatt wind farms. Over a few years, Zilkha Renewable had gained confidence and competence. With Skelly as the chief development officer, the company had doubled in size and then doubled again. Before long, it was building 300- and 400-megawatt wind farms. Costs had risen significantly, and so had Zilkha's stress level.

Zilkha financed these larger wind farms from his balance sheet. He took out his checkbook and paid for land leases and environmental studies, salaries, and turbine down payments. He didn't have a billion-dollar line of credit from a bank. He hadn't taken on debt to pay for these wind farms.

When he created Zilkha Renewable Energy, he wanted to have no more than $100 million sunk into developing projects at any one time. Maybe a bit more, depending on how the projects looked. But the basic idea was that he would open up the Zilkhas' checkbooks to get the projects started. Some would wither or fall apart. But others would develop momentum. The land-use approvals would be signed and deals struck to sell the electricity. At that point, Zilkha could sell off the wind farm or bring in a partner. Either way, funds would flow back to the Zilkhas and their checking accounts. They could then recycle that money back into the business.

Michael Zilkha wanted to limit the amount of money he had at risk. He called it being "pregnant." By the summer of 2004, the company was hitting its stride. Michael Skelly and his team were getting skillful at finding and developing wind farm prospects. Negotiations over Blue Canyon were moving ahead. Skelly and others could finally anticipate the beginning of construction—and the big expenditures that would be required. At the same time, work was set to begin the next year on the largest wind farm east of the Mississippi River. Zilkha owned half of it and would be on the hook for half of its $380 million budget. And there were other projects in the works—one in California and another in Minnesota.

Skelly and the company's financial team were delivering the same message: We need more money. Skelly crunched the numbers, as did Jayshree Desai, the chief financial officer, and Joe Romano, the chief executive. The Zilkhas were going to need to spend somewhere in the neighborhood of $500 million in 2005. The little company was about to hit a growth spurt. The monthly pilgrimages to Michael Zilkha's office to ask him to write a check were going to get a little more vexing.

"As long as we could write the checks, we wrote the checks. Because we believed in it," said Michael Zilkha. But at a certain point "we couldn't afford it anymore." He was too pregnant. So he went to Wall Street for help.

In the fall of 2004, Michael Zilkha went to New York. His first stop was Goldman Sachs's headquarters at 85 Broad Street near the southern tip of Manhattan, a precast concrete tower that an architectural critic once said was "one of the most forgettable tall buildings in New York." The relationship between the Zilkha family and Goldman Sachs dated back generations. When the Zilkhas decided it was time to sell their oil and gas company, Goldman brokered the deal. The price was more than $1 billion.

Michael Zilkha explained his situation to the Goldman bankers. The company was growing and so was the amount of cash needed. Zilkha Renewable had outgrown being a family-owned business.

In a follow-up dinner meeting in December, Goldman bankers gave Zilkha a choice. If he wanted to hire Goldman to put his company up for sale and find a buyer, they were happy to do that. Or, if he was interested, he could sell his company to Goldman Sachs itself.

The Wall Street firm was interested in wind. It was growing and had strong support from the states and the federal government. There were few renewable energy companies in the industry with strong balance sheets. Goldman Sachs saw an opportunity to buy the company, impose some financial controls, and help it grow.

Not long after the dinner, Skelly, Desai, and Rick Winsor, the Canadian engineer who focused on project details, flew to New York for a meeting where Goldman Sachs bankers would grill them. Investment bank employees escorted them past security and up to the executive floor. They deposited the Zilkha team in a conference room. As they waited, a tall, well-attired man with glasses stormed into the room with his overcoat still on. "I'm lost," he said. "Who are you guys?" They chitchatted for a few minutes. Then he introduced himself: Hank Paulson, the chief executive of Goldman Sachs.

"He wasn't lost. He knew exactly where he was," said Winsor. "He came in and checked us out before the rest of the troops." Whatever test he had applied, the Zilkha team passed.

Within a couple of months, they struck a deal. The Zilkhas kept one-eighth of the business, but sold the rest to Goldman Sachs. They would keep the company in Houston. Much of the culture remained: the open-floor layout and open lines of communication. Michael Zilkha repatriated some of the rock 'n' roll photographs to the family offices upstairs, although many remained.

The new owners didn't want to lose key personnel. Keeping Skelly was a top priority. "Michael was the animating spirit. He was the soul of the company," said Neil Auerbach, a Goldman Sachs partner who helped broker the sale.

In March 2005, Goldman Sachs announced the deal. Zilkha had almost 4,000 megawatts worth of wind energy projects in various stages of development across twelve states. Neither company ever disclosed how much Goldman bought the company for, but documents and interviews suggest it was several hundred million dollars. The way the deal was structured, the Zilkhas' payment didn't come all at once. It came over time as various projects were completed. That made it essential to keep Skelly, the chief development officer, productive. If anything, the new regime unleashed him. He needed to plant more wind turbines in the earth.

At the time, Wall Street was discovering the wind industry and the tax benefits it offered. It was a marriage that would help propel the growth of the renewable energy industry in the United States. The biggest attraction for Wall Street wasn't an opportunity to paint itself green. It was the tax credits for wind production that allowed companies such as Goldman Sachs, and its clients, to lower their tax bill.

The U.S. government has a long history of using tax policy to promote energy sources. Beginning in 1916, Congress wrote breaks for oil and natural gas developers into the tax code. Companies could write off various costs of drilling wells. The result of these and other breaks, some of which still exist, reduced production costs, attracting more investment into the industry. Between 1920 and 1970, the discovery and use of fossil fuels rose exponentially.

In the 1970s, power companies received massive federal income tax subsidies to build more nuclear power plants. In 1979, a noted resource economist estimated that 35 percent of the cost of generating nuclear power was paid in tax subsidies. The industry also benefited from a government-imposed limit of its liability in case of a nuclear accident. Without the last provision, enacted in 1957, no one would build power plants because an accident and resulting lawsuits would bankrupt any company. Without all of this aid, there wouldn't be any nuclear power in the U.S.

By 1980, the government changed the tax code to encourage more natural gas exploration. Facing a potential gas shortage, it gave tax credits to companies that produced gas in unconventional places, such as coalbed seams and shale formations. These credits expired in 1992, but they "helped spawn and push into use an entire new set of exploration, completion and production technologies founded on improved understanding of unconventional gas reservoirs," two economists wrote in the *Oil & Gas Journal*. The technologies continued to evolve and improve. The result was fracking and a fossil fuel energy boom.

As this tax credit expired, Congress introduced something new to support the nascent wind industry: the production tax credit in the Energy Policy Act of 1992. Every kilowatt hour of electricity produced by a wind farm would produce a 1.5-cent tax credit that could offset federal tax. The amount of the tax credit rose with inflation. By 2006, the Blue Canyon wind farm was generating about $15 million in annual tax credits. Goldman could use those tax credits to offset its substantial tax bill. In these pre–Great Recession days, Goldman was doing quite well and had a large appetite for tax credits to offset their billion-dollar profits.

Ken Pontarelli, a Goldman Sachs managing director who helped oversee the deal, says the company was impressed with the Zilkha team. In Skelly, he says, "we had a world-class developer." The Zilkha crew was good at going out and leasing windy sites where power prices were high. Goldman could sell the power output to a utility and keep the renewable energy tax credits for itself.

Goldman gave the company a new name to reflect this forward-looking stance: Horizon Wind Energy. Under Goldman, the company transformed from a developer of wind—a company that would scout out locations, build wind farms, and then sell them off—to an independent power company. It could build wind farms and hold on to them for years, operating the turbines and selling the electricity.

The newly named company went on a turbine-shopping spree. Two Goldman executives and Rick Winsor flew to Europe, stopping in Denmark, Germany, and Spain. Winsor said they bought $2 to $3 billion worth of turbines. "It was something the Zilkhas clearly couldn't have done," he says. "Now we could translate our opportunities into executable deals because we had the turbines."

European turbine manufacturers had been cautious about entering the U.S. market. Every few years, Congress would let the tax subsidy lapse and then reinstate it later. The wind market would roar to life when there was a subsidy, then collapse when it went away like a pinpricked balloon.

The U.S. used about a quarter of all the electricity in the world, but its renewable energy market was tiny. Companies wanted to break into it, but were reluctant to work with small developers without a track record. But as Winsor remembers it, a Goldman Sachs–backed company ready to spend a couple of billion dollars instilled confidence.

By the end of the year, contracts were signed. There was a $700 million deal with Gamesa to purchase 600 megawatts' worth of turbines. A month later, almost exactly a year after Goldman entered into negotiations to buy Michael Zilkha's company, it placed another order for 600 megawatts, with an option for another 200 megawatts. Meanwhile, Skelly was developing wind farms in Indiana, Oregon, Texas, and Washington.

Goldman Sachs could bring a checkbook to purchase wind turbines that dwarfed the Zilkhas'. And they could bring financial structure to a company that was essentially an overgrown start-up. But they needed the zeal and commitment of Skelly's team of developers.

Pontarelli said the mainspring was Skelly. "This is a guy who rides his bike to work every day, you know. He loves what he does. He is like a millennial generation person before there were millennials. He believes in what he is doing. And when you believe, when you are going out into upstate New York, or Illinois, or Iowa, people can smell that. And they know these are good people and they are going to do what they say they are going to do."

———————

Josh Magee joined the company in the last few months before Goldman bought it. He was one of Skelly's typical hires: youthful, smart, and passionate about the environment. He had graduated a year earlier from Stanford University, where he had studied renewable energy, and then received a fellowship to get training in environmental political organizing. He believed the wind industry was maturing, and he wanted to be on the side of building renewable energy not getting petitions signed. He sent out his résumé, which was thin at the time, to every renewable energy company he could find in the U.S. He got one nibble, from Skelly, who invited him to come to Houston for an interview. They talked for a while and Skelly hired Magee on the spot. In one of his first days, Skelly asked him to attend a meeting about a project in Maine. He might learn something about how projects are developed.

The meeting was small, just Skelly, Winsor, and Desai. They were discussing a problem in Maine. To be more exact, Magee said, Skelly was gesticulating in front of a large map of the state, moving back and forth while brainstorming out loud as the financial duo watched.

The project, called Aroostook, was remote. A local partner wanted to erect turbines in the potato fields around Frenchville, Maine. If you imagine a map of the United States, there are two bumps at the top of Maine. The eastern bump is the potato field area, about as far from the rest of New England as possible. It is closer to Dublin, Ireland, than San Diego, California. The wind in the Saint John River Valley was good, but getting the blades up there was going to be a logistical challenge.

Skelly had an uncapped marker and drew out options for the transmission lines. One would run for a couple of hundred miles south from the potato fields, connecting to the New England grid. It was the shortest route, but would require disturbing pristine habitats.

Desai and Winsor looked at the map and the lines Skelly had drawn. "I don't see it working," Winsor said. Skelly stared at the map a bit more and then took out a marker and drew a line straight up into Quebec. "That's how we'll do it," he announced. The power would go up into Canada and then follow a counterclockwise circle back into New England. That wouldn't be cheap, so the wind project had to be very large to spread the costs of the wires across a large number of megawatts. To drive down the cost of electricity, the wind farm would need to be bigger than originally planned. Skelly talked about a 500-megawatt wind farm with 250 turbines in the potato fields. It would be the biggest east of the Mississippi. Skelly broke up the meeting. Magee was impressed. "It was 'don't tell me it can't be done.' Tell about the constraints and let's throw ideas around. Let's make this a reality even if it seems far-fetched."

Magee said he thought Skelly and Winsor had "an interesting and beneficial dynamic. Skelly was the dreamer and Rick was the person who said 'here is the dream, how do we get it done in a way that everything works and we don't get sued and we make as much money as possible.'"

In September 2007, Horizon grew the project again, hoping that by getting bigger, it could spread out the cost of wires and make the numbers work. Growing the project required building an 800-megawatt project in a part of the country with only 36,000 customer accounts. The only way the numbers added up was to connect northern Maine to the rest of New England, and that would require $625 million in new power lines and substations. Since the power would serve New England, the cost of building the new wires should be spread across all the region's ratepayers, they argued.

The rest of New England wasn't interested in Maine's wind. The request is "astonishingly audacious," countered the Connecticut Office of Consumer Counsel and the state's Department of Public Utility Control. Other problems arose, including concern that adding so much wind in a distant corner of the grid could cause instability. The project stalled and ultimately died.

By this point, Magee had moved to Albany for Horizon to help develop the Maple Ridge Wind Farm. Working for Skelly, he said, meant always hustling. You didn't want to let him down. His energy was infectious. With Goldman's money and turbines being manufactured, there was always more to be done. Magee tried to develop other wind farms in upstate New York, but kept running into community opposition. He could tell how much locals were going to fight him based on the number of vacation homes. "If the number of vacation homes in that town were high, there was going to be a high level of opposition there," he says. "If it was low, there would be low opposition. That was true town to town across upstate New York."

Magee only stayed at Horizon for a couple of years. He left because he was twenty-six years old and living out of motels in places such as Batavia, New York. He wanted a better quality of life. He stayed in renewable energy, mostly writing research reports. He said that Goldman Sachs's purchase of Zilkha Renewable helped change the perception of the wind industry in the United States. The investment bank had an air of invincibility about it, at least until the real estate bust humbled Wall Street banks. Goldman's entry into wind implied there was money to be made in renewables.

When Skelly hired Magee, the wind industry was still a lot of small, poorly funded developers. By the time Magee left Horizon, it was to write analyst reports on the state of the wind industry for institutional money managers who wanted to invest in it. Wind had become a big business.

Magee went to work for Emerging Energy Research, which provided market intelligence and insights on the growing industry. By

2007, he wrote in a report, Horizon had become the second-largest wind developer in the country. It was the third-largest owner of wind farms. The tiny company, which had three employees, including Skelly, in 1998 was a major wind power producer in the fast-growing wind industry. In 2006 and 2007, the U.S. added more wind power than any country in the world.

The decision to make Aroostook even bigger was one of the last that Skelly made working for the company he helped create. In March 2007, Goldman Sachs sold the company to Energias de Portugal SA for $2.15 billion plus debt of about $180 million. This was a healthy profit, about $900 million, factoring in capital Goldman had put into Aroostook since buying it from the Zilkhas.

EDP, as the company is known, was a giant in global renewable energy. It wanted to get in on the biggest, fastest-growing market in the world: the United States. Buying Horizon gave them a pipeline of projects and a team that knew how to operate on the local terrain. And even though EDP didn't have a large U.S. tax to offset, there was a market emerging to sell the tax credits to banks and insurance companies.

Gabriel Alonso, an EDP executive who joined Horizon as soon as the sale closed in July, said acquiring Horizon was like "buying a Formula One race car." Alonso liked Skelly and recognized he was the heart of the company, even though he wasn't the CEO. "Let's be clear. He was *only* the chief development officer? That was the company. The company was doing development," he said.

EDP's approach differed from that of Zilkha Renewable and Goldman Sachs. Zilkha had built wind farms and then sold them. There was no long-term strategy of managing the wind farms. Goldman Sachs never intended to own the company for a long time. When EDP took over, Horizon's asset management department consisted of a single person. EDP planned to build wind farms and own them for decades or longer.

Under Skelly, Zilkha Renewable/Horizon had been a hard-charging company. It needed to move aggressively to grab opportunities. It was built to grow. EDP felt they would need to reorient it as a company with a slower metabolism focused on the long term.

The Portuguese executives liked Skelly and wanted him to stay. But Skelly was ready to be chief executive. EDP wanted its own person to run its new North American operations, which it had just spent considerable money to purchase.

Skelly decided it was time for something different. On the first day of December 2007, he quit Horizon and left the wind business. Eight days later, he filed paperwork with the Federal Election Commission.

He would run for Congress in Texas's Seventh District—a Republican stronghold. He filed as a Democrat. His wife, Anne Whitlock, was listed as the treasurer.

"It was a lot more interesting than other stuff I could have done," he said.

// **8** //

"Tell Me If I Sound Like a Liberal"

I n his first days as a political hopeful in December 2007, before he officially entered the race, Michael Skelly paid for a poll. He was comfortable being the underdog. He just wanted to know how much of an underdog he was. The answer that came back from several days of calls to hundreds of potential voters in the Seventh Congressional District of Texas: Skelly had a slim chance. Voters didn't particularly like the incumbent and they thought the country was moving in the wrong direction.

"Once voters hear about Skelly's background as a successful wind energy businessman and his message of change," the polling concluded, "Skelly closes the gap and draws even with the incumbent." This diagnosis was good enough for Skelly.

He issued a press release announcing he was running and rented a campaign headquarters, a second-floor space above a café on Kirby Street near his house. He then turned his attention to meeting the voters and building a staff. He hired a campaign manager who scheduled long days all week long, except for large chunks of Saturday. Skelly reserved those for watching his children's soccer games. He devoted the rest of the week to figuring out how to get affluent residents of West Houston to do something they hadn't done since Skelly was in kindergarten: vote for a Democrat. The district's voters had sent a Republican to Washington, since 1967, the year an oilman named George H. W. Bush won the seat.

Skelly had a long-standing interest in government. At Patrick Henry High School in Roanoke, he was on the executive council of the student government. At Notre Dame, he wrote opinion pieces for the student newspaper about nuclear weapons and divesting from South Africa. In one column from March 1983, he urged his fellow students to slough off complacency. "We should appreciate America for what it is, but not lose sight of what it could be," he wrote. What interested him about politics was not the uplifting oration of a campaign speech. He liked policy. He believed that smart, well-crafted government policy could help people. When Skelly first talked about running for Congress with his wife, Anne, she was not surprised. "In the back of my mind," she said, "I thought it was inevitable."

Candidate Skelly had some clear strengths. He was smart and articulate. He was personable and thanks to the sale of Horizon to EDP, he was wealthy enough to go without a paycheck for a few months and lend his campaign funds as needed. And he had a backstory that might appeal to voters. He could talk about growing up in an immigrant family and giving back to the country that had given his family a chance.

The incumbent in the Seventh District was a conservative Republican named John Culberson. His main political passion was to bring home to Houston as much federal money as possible to expand the interstate that connected downtown Houston to its western suburbs. He scoffed at any alternative form of transportation. Spending money on a light rail in Houston, he once said, was "like putting a water hose down a gopher hole."

In his early fifties, Culberson had wide shoulders and a thick neck and looked able to suit up as a football lineman. A staunch conservative, he disliked government. But by 2008, he had spent his entire adult life in it. After graduating from college in Dallas, he went to work for his father's political consulting business. The elder Culberson was a graphic artist who designed brochures, billboards, bumper stickers, and mailers for candidates.

After about five years with his father, Culberson was elected to the Texas House of Representatives in 1986, only five years out of college and a few months shy of his thirtieth birthday. He did not distinguish himself in Austin. Around the capitol, he is remembered for his role in ending years of oversight of state prisons by a federal judge and for holding himself out as the smartest guy in the room. His colleagues in the state legislature, both Republican and Democrat, weren't upset when, after fourteen years in Austin, he won an election for an open seat in Congress and left for Washington.

In Congress, Culberson attempted to ingratiate himself with his constituents by talking about his desire to disembowel the Internal Revenue Service. "I want to do to the IRS what Rome did to Carthage, tear it down stone by stone and sow salt in the furrows," he said at a town hall meeting in 2001. He was an antigovernment government official and a Tea Party conservative before there was a Tea Party. Despite his attempts to discredit and raze government institutions, Culberson wasn't above bringing home pork. His constituents wanted to cut the federal government down to size, but they also wanted it to help pay to widen their roads.

Culberson had represented many voters of the Seventh District for more than two decades by the time Skelly decided to run. Culberson knew the precinct chairs. He knew the issues that stirred the voters and the issues that comforted them, especially the conservative suburban voters in the western half of the district. That core constituency knew him and voted for him reliably.

Voters had returned Culberson to Congress four times after first electing him in 2000. Culberson secured a seat on the all-important Appropriations Committee, where he helped control the government's purse strings. If he ever needed to fill his campaign bank account, political action committees were more than willing to help out.

––––––––––––

The demographics were against Skelly. In 2008, the Seventh District was what political operatives called an R+16. That meant, all else being equal, the voters of the district would favor a Republican candidate over a Democratic candidate by a 58 percent to 42 percent margin. Skelly's political consultants convinced him that he could close that gap by tapping into what his pollsters said was a significant dissatisfaction with President George W. Bush, the direction of the country, and Culberson. "Once voters hear positive biographies of both Culberson and Michael Skelly, the race draws to a dead heat," a memo to the candidate confidently asserted.

Skelly looked at the demographics and, as was his habit, saw what could be, not what had been. Why wouldn't the voters get excited about a newly minted energy millionaire and take a chance on a political newcomer?

The district had the highest median family income in Texas—and one of the highest in the United States. It started around Rice University and then headed west, scooping up many of Houston's older, wealthy neighborhoods. In addition to former President Bush, who lived in the district for decades, other Houston royalty lived along the green lawns of the River Oaks Country Club and other well-to-do neighborhoods, including corporate CEOs, basketball legend Clyde Drexler, and television pastor Joel Osteen. Enron founder and chief executive Ken Lay's 5,500-square-foot house—and his thirty-third-floor condominium—were both in the district.

It wasn't a completely hopeless district for a Democrat such as Skelly. Some of the socially conservative voters on the western side of the district had been peeled off and attached to another district in a recent redistricting. This left behind what a political journalist said was "reliably Republican, but more business-minded territory."

Republican voters left behind, especially those inside Houston's loop, tended to be more pro-choice and pro–gay rights than the rest of the party. Perhaps they would be open to a Democrat, if he could demonstrate a commitment to capitalism—Houston's unofficial religion. Skelly certainly had the right résumé. In his campaign announcement, he positioned himself as a "successful renewable energy entrepreneur." He was a wind guy, but made sure to point out that he was in the energy business, someone very much like the people of West Houston. Skelly pitched himself to voters as someone who had created jobs and grown a business.

Skelly rolled through the primary unopposed. To finance television advertisements in general, he started calling friends and business associates. "He had a tremendous Rolodex. And he wasn't twisting arms," says Bill Kelly, his first campaign manager. "Not only did he know a lot of people, the people he knew genuinely liked him and wanted to help."

Culberson sized up Skelly and decided to try to attack his connection to renewable energy. He wanted to paint a picture of Skelly as a wind farmer grown fat on government subsidies. The oil industry had turned Houston from a swamp into a wealthy metropolis; Culberson portrayed Skelly as a man whose business was out to kill fossil fuels. "These people in the wind energy business have made their fortunes because they are subsidized by you and me," he said in a video chat with constituents. "And the Democrats passed a big energy bill late last year that jacked up taxes on the oil and gas industry by about $14 billion and then handed the money over to the wind energy folks and other industries. That's just dead wrong."

By June 2008, a few months into the campaign, Skelly had begun to draw national attention as a fundraising machine. He had brought in about $800,000, the most of any Democratic challenger for a House seat. The local and national media began paying attention.

Skelly was outraising Culberson. He tapped into the deep pockets of Houston's social, energy, legal, and financial elites. Lawyers at the city's big corporate law firms, Baker Botts LLP and Vinson & Elkins, were major contributors. But so was Rusty Hardin, a flamboyant local defense lawyer who represented former baseball pitcher Roger Clemens against charges of lying about steroid use before Congress. A violinist with the Houston Symphony contributed $364. The proprietor of McGonigel's Mucky Duck, a beloved venue for Texas guitar troubadours, gave $107. Money came from Lynn Wyatt, whom *Vanity Fair* has described as "Texas's ur-socialite," and from Lois C. Chiles, an actress best known for playing Bond girl Dr. Holly Goodhead in *Moonraker*. Labor political action committees and West Coast tech entrepreneurs contributed. A limo driver and a biology teacher were among the nearly two thousand donors. So were more than a dozen contributors who listed Goldman Sachs as their employer. Matthew Simmons, a former energy advisor to President George W. Bush and éminence grise in Houston energy banking circles, gave $4,600, the maximum allowed in the primary and general election. He wasn't the only oil industry leader. Doug Foshee, chief executive of El Paso Corporation, and James Woods, the retired CEO and Chairman of oilfield service firm Baker Hughes both backed Skelly.

Pat Wood III, a longtime associate of President George W. Bush and former chairman of the Federal Energy Regulatory Commission, threw a "Republicans for Skelly" event at his home in Houston. He had met Skelly a year earlier and liked his message that Houston should be in favor of all types of energy—oil, gas, wind, solar, whatever. "I had no love for the incumbent. Still don't," Wood said, so when Skelly approached him he was happy to help. Fayez Sarofim, a Houston money manager and skilled investor who donated generously and reliably to the GOP, gave Skelly $4,600.

Political observers were impressed that the neophyte was managing to outraise the incumbent. For every $1 that Culberson raised from individuals, Skelly brought in $1.70. Culberson closed that gap by bringing in thousands from PACs representing shipping and energy companies, defense contractors and pharmaceuticals makers. A few political observers started to think that maybe, just maybe, Skelly could overcome the demographic long odds and steal the seat. Political analyst Stuart Rothenberg provided one of the dissenting voices. "What about Michael Skelly? He is smart and would be an engaging dinner companion. If Skelly were running in a competitive district, I'd think he'd have a good shot. But he isn't," he wrote during the summer. The Seventh District is a "nightmare for any Democrat." Could Republicans beat Barney Frank in Boston, even in an off year for Democrats? Not likely. Rothenberg concluded that even in an off year for Republicans, Skelly didn't have a shot.

Skelly didn't show any signs of doubt. He was an optimist and had convinced himself that he could win. His political consultants urged him on. Internal polling showed the sixteen-point gap narrowing.

By midsummer, it was clear Skelly was the strongest candidate to ever challenge Culberson. But the incumbent had what he thought was a secret weapon. He had heard that Skelly had hired a military helicopter from the Sandinista air force to move towers for the aerial tram in Costa Rica. If necessary, he could make sure it leaked that Skelly had done business with Central American communists. In a conservative Houston congressional district, just a whiff of association with communists, an innuendo of fraternization, and the election would have been over.

Of course, there was an explanation. He had done business with the air force of a sovereign government, using the only helicopters in Central America capable of moving the giant towers. And at the time, left-wing Sandinista revolutionaries weren't leading Nicaragua. But the rule in politics is that if you're explaining, you're losing. And if Skelly had to explain why he had hired the Fuerza Aérea Sandinista, he wouldn't get the chance to measure his new congressional office for drapes.

During the summer, Skelly heard that Culberson knew. "He is

putting the word out in D.C. that he can blow me out of the water whenever he wants," he said. Skelly's consultants wanted to do a poll to see how voters would react. Skelly thought this idea was stupid. "We are going to do a poll to see if people are going to vote for a guy who did business with communist terrorists? I don't think we need to do a poll about that," he said. A political bomb now hung over the campaign.

Skelly called Joanne Herring, a Houstonian who was uniquely qualified to help. She had experience with Russian-made helicopters and with getting Republican congressmen to do her biddng. In her late seventies, Herring remained a force of nature—a wealthy, twice-divorced, once-widowed woman dubbed the "Queen of Texas"—who was regularly referred to as a blond bombshell. Skelly called the bombshell to defuse the bomb.

As a younger woman, Herring was a Houston hostess known for her commitment to throwing a memorable social event. *Life* magazine sent a reporter and photographer to her thirtieth birthday party. A photograph shows the young hostess, wearing flowers in her hair, long earrings, and a sleeveless white dress, handing out lots to see which partygoer would be burned at the stake. Not literally, of course. *Texas Monthly* would later describe it as a "Roman orgy theme party."[1]

Her second marriage, to a natural gas tycoon, drew her into the political world. Sally Quinn, chronicler of Washington, D.C.'s social scene, wrote a two-part series about her in 1978 that ran to more than seven thousand words. Herring and her husband "are now pretty much regulars on the Washington party scene, one night at a party for the king of Morocco, another evening dining at Henry Kissinger's, another at a farewell party for the Pakistani ambassador," Quinn wrote.

In the 1980s, not long after the death of her energy CEO husband, Herring started dating a scandal-prone, underachieving congressman from East Texas named Charlie Wilson. She introduced him to Pakistan and then to Afghanistan. Herring felt that the United States needed to support the Afghan fighters who were engaging the Soviet

Union's army. Over the course of a few years, Herring and her congressional beau secured secret funding for Afghan fighters. She once accompanied Wilson to the Paris Air Show in search of portable anti-aircraft weapons for Afghan fighters. It was there that she began to become informed about Russian helicopters.

The partnership of a socialite and a lawmaker became a book. Then Aaron Sorkin turned the story of the odd partnership into a screenplay. The resulting movie, *Charlie Wilson's War*, opened in the final days of 2007 and made Herring into a bit of a cult figure in Houston. It didn't hurt to have Julia Roberts play her on-screen, with a blond, architecturally impressive hairdo. Herring insists she was not the trollop portrayed in the movie, but otherwise enjoyed the bright lights of fame that illuminated her life in River Oaks, in the heart of the Seventh District.

Skelly called her for help only a few months after Herring walked a red carpet at the movie's Houston premiere in a red lace fishtail gown. She was intrigued. As he explained his situation, she realized that in a past visit to Costa Rica, she had taken the aerial tram through the rain forest. "I thought it was wonderful. I loved what he did and I thought it was terribly good for the area," she said. Skelly's ability to build the tram impressed her. As far as she was concerned, Skelly had hired the Nicaraguan helicopters to solve a problem. "You had to be practical or you are never going to get anything done," she said.

She was upset that Culberson was threatening to use Skelly's accomplishment against him. When Skelly asked her to call Culberson, she readily agreed. It was wrong to make a political issue out of this, she felt. If Skelly had helped Nicaragua beat their swords into plowshares, it would be unchristian and un-Texan to be penalized for it.

She called Culberson and spoke to him in a sweet but determined voice. She explained to him that she had been fighting communists for her entire life, since she joined the John Birch Society in her twenties. Damning Skelly by association with the Sandinistas wasn't the right thing to do, she said. He helped bring jobs to Central America. Don't make this a campaign issue, she said. She recalls Culberson agreeing not to use it. "He could have blown it up and said the Sandinistas are our enemy, but he didn't. He deserves some credit," she said.

It is not clear if Culberson ever planned to use the information. Perhaps he intended to keep it in his pocket, an insurance policy in case the election took a turn. Whatever the plan, Joanne Herring's charm worked well enough that word got back to the Skelly campaign that the issue wouldn't surface. There was never a peep about it on the campaign trail. The issue defused, Skelly and his campaign turned their attention to introducing him to the voters.

Starting in the summer, Skelly's television commercials ran in Houston. Purchasing airtime wasn't cheap. Houston was the fourth largest media market in the country. Despite his fundraising prowess, Skelly needed more money. In late June, he reported that he had lent his campaign $200,000.

The first television ad to hit the airwaves was a warm-and-fuzzy introduction to Skelly. In the spot, his kids do most of the talking. One of his sons talked about how Skelly thinks Congress can do more to solve the energy crisis, like using more renewable energy because the oil won't last forever. "He talks about that all the time," said his daughter, a young teenager at the time, who turned out to be the giggly star of the commercial. Skelly barely said a word.

To be palatable to centrist Republicans, the polling suggested that Skelly needed to run as a pro-energy Texan and moderate Democrat. During the campaign, Culberson tried to define Skelly as a liberal in lockstep with out-of-state Democrats. The Republican Party of Texas sent out a mailer to voters that showed Skelly's face next to Barney Frank's, Nancy Pelosi's, and Tom Daschle's. "They're four peas in a pod," it read. "Washington has way too many liberal elitists. Don't elect another one." The mailer concluded that Skelly was trying to hide that he was a left-leaning Democrat. "Vote NO on Liberal Michael Skelly," it read. "Michael Skelly can't be trusted." In speeches and campaigning, Culberson would return to this theme over and over. Skelly was an archliberal in disguise, ready to show his true colors the second he was sworn in.

To combat this, Skelly went door-to-door in the district. He was good at it. Get him one-on-one with a voter, on the doorstep, and he could fall into a conversation with ease, building rapport and trust. His staff sometimes had to pull him away from a house and remind him there was the rest of the block and the next block to cover. Skelly didn't mind the triple-digit heat and round-the-clock humidity. He enjoyed meeting people.

But changing minds one house at a time was a hard way to make up the needed votes. As the summer wore on, Skelly's best chance was for Culberson to make a gaffe of epic proportions. Culberson didn't cooperate. But he did put his foot in his mouth once. He was speaking to an online town hall when he said, "NASA is a waste of money." This did not go down well in a city that had played such a pivotal role in the space race. He apologized and the mini-controversy passed.

———————

Hurricane Ike hit the tip of Galveston Island, less than fifty miles south of downtown Houston, in the early hours of September 13, 2008. It had sustained winds of 110 miles per hour and pushed a storm surge ahead of it. Its landfall was unsettlingly reminiscent of another hurricane, a century earlier, which had left thousands dead.

In the days before Ike hit the Texas coast, first meteorologists and emergency preparedness officials and then the entire metropolitan area of Houston kept a close eye on it. The projected path took the storm from the cool Atlantic waters into the bathtub-warm Gulf of Mexico. Computer models predicted it would make landfall along the upper Texas coast, right where Houston was located. The previous three years had been filled with hurricanes. Gustav, Dolly, and Rita, along with Tropical Storm Eduardo, had all barreled into the upper Texas coast. And Hurricane Katrina had hit just to the east with biblical destruction, leaving a deluged New Orleans in its wake. Who could blame Houston for focusing for days on advanced weather reports and barometric pressure readings?

After Ike smashed into Galveston, it entered Galveston Bay and made a second landfall, this time on the mainland, just east of Houston. In Harris County, home to Houston, nearly one in twelve homes sustained damage. The storm's intense air pressure blew out thousands of windows in the downtown skyscrapers. Shards of glass covered the streets. Trees were uprooted across the city, tearing down power lines as they fell. More than three million people lost power. By the time Ike moved northward and residents emerged from their homes, they were greeted with oppressive humidity, late summer heat, and no power. A political campaign was the last thing on most people's minds.

When Skelly arrived at the campaign headquarters after Ike, he opened the door to a suffocating oven. The electricity was out and without air-conditioning, in the late Texas summer, the office was pretty much uninhabitable. The campaign moved across the street to the rented apartment of Dylan Loewe, a recent Stanford graduate who had taken over as campaign manager before the storm. His apartment was in a relatively new complex. The power lines into it were underground and protected from storm-ravaged tree limbs. He had cool air and an internet connection. Staff and volunteers squeezed into the space. Skelly often took over the bedroom, lying down on the bed to make fundraising calls.

Meanwhile, a new batch of internal polling showed that Skelly was closing in on Culberson. A double-digit deficit earlier in the year had been cut to five points. Loewe determined that his role was to keep Skelly and volunteers motivated. "I thought it was important to tell everyone including Skelly that this was in reach," he remembered.

After the hurricane passed through, Skelly filed another campaign finance report. He had lent his campaign another $750,000 to keep his television ads on the air for the next few weeks. Skelly was all-in. His campaign ended up spending more than $1.1 million on 1,700 ads on network television and another $275,000 on cable television. Culberson spent $900,000.

As the campaign entered its final weeks, Skelly positioned himself as more fiscally conservative than Culberson. In one ad, he's sitting at a table in an office wearing a button-down shirt and blue tie, and says: "John Culberson says I'm a liberal. Well, tell me if I sound like a liberal. I turned a two-person start-up business into a multibillion-dollar wind energy company. I support drilling for oil and renewable energy to make us energy independent. I agree with John McCain that we have to get rid of the wasteful earmarks that John Culberson supports."

Jennifer Naedler, Culberson's campaign treasurer, says Skelly's tactics were surprising. "He ran to the right of the congressman," she said. The Seventh District's pool of liberal votes was shallow. Skelly's tactic was to wade into Culberson's turf and cast about for votes among conservative voters.

In the final days of the campaign, Skelly released his only negative advertisement. It was sloppy, with bad graphics, in sharp contrast to his other ads. It mocked Culberson for running up a $10 trillion national debt, allowing the subprime mortgage crisis, and being against earmark reform that contributed to a culture of wasteful spending in Congress. Naedler blamed the campaign's "unnecessarily nasty" turn on Skelly's out-of-town hired guns. "It felt overly personal. And that's not Michael Skelly, those are his consultants who talked him into it," she said.

Late in the campaign, the Texas Republican Party sent out a new mailer. "Private property rights? Not if Michael Skelly has his say," the mailer declared. "Michael Skelly thinks the government should be able to confiscate our private land. . . . Why? To build more power transmissions."

The terminology is clunky, but the political objective is clear to see. Wind farms require transmission lines, and building new lines can require the use of eminent domain. Government confiscation of private land is never a popular idea. Culberson's Republican consultants had spotted a weakness.

Of course, this line of attack only works if you don't dig too deeply. Michael Skelly had barely ever needed to use eminent domain to build a wind farm. But Culberson had campaigned on his ability to expand the interstate in West Houston. It was his pet project and his main issue. It required considerable use of eminent domain and the eviction of dozens of businesses and people.

Culberson made sure that there were several ribbon-cutting ceremonies for various stages of the project during the campaign. In October, just before the election, Culberson smiled for photographers, holding a golden hammer while hundreds of balloons were released beneath a multilevel interchange to celebrate the official opening of another stage of the interstate expansion.

The Katy Freeway, as that stretch of Interstate 10 is known, was opened in 1968 as a rural highway heading west out of Houston toward nearby Katy. It was designed to carry 79,200 vehicles a day. By the 1990s, it was one of the most congested parts of the federal interstate system. The traffic on it was nearly three times its design capacity. There were traffic slowdowns eleven hours a day. The weight of all the vehicles punished the road. Maintenance cost $200,000 per mile every year. And costs were rising. The highway ran through the heart of the Culberson's congressional district. Culberson, in the official biography on his website, talks about his belief in limited government. But in the same document, he brags about bringing federal money to the district to expand the Katy Freeway. This $2.8 billion project used taxpayer money to speed up the commute for his constituents in the western suburbs to their jobs downtown. Culberson said opening up the traffic-clogged road was like "giving West Houston a quadruple bypass."

What he doesn't talk about is that $480 million of the cost went to buy private property in order to widen the freeway and provide nine lanes (two toll lanes, three frontage road lanes, and four main roadway lanes) in each direction. To expedite the Katy Freeway expansion, Texas hired a contractor who subsequently acquired 253 parcels of land.

More than six hundred residents and business owners were evicted from their property. Eminent domain was required. One business, a restaurant called Caffe Ribs, was offered $7.4 million for a 7.5-acre tract. It didn't like the eviction or the payment, and the company spent the next nine years appealing the decision all the way to the Supreme Court of Texas, which remanded it back to trial court.

A local grocery chain, Fiesta Mart, closed one of its most popular stores when the state took 40 percent of its parking lot. The outdoor retailer REI moved when it lost part of its parking. A local photograph-printing store had to relocate. The move "definitely hurt us," owner George Lamb told the *Houston Chronicle*. "If it weren't for our customers being so loyal to us, we'd have been out of business a long time ago."

A political campaign isn't the place for a nuanced conversation about eminent domain. Government has an awesome power to take land away from private owners when there is a public, common good at stake. Culberson was for eminent domain when it was used to widen a highway and speed up his constituents' commute. He was against it when it was used to build transmission lines.

———

On November 4, 2008, Skelly spent the day in a crush of last-minute campaigning. His belief that he had closed the gap kept him going. That is what the polls suggested and what his campaign manager had told him.

The Skelly campaign held a vote-counting party downtown at an old Houston business club. His wife, Anne, said that when she arrived at the club and got off the elevator, he ushered her into a backroom. "It doesn't look good," he said. She remembers the look on his face. She hadn't seen it before. He looked numb. "He's an optimist by nature. He only sees the glass half full," she said.

The early voting results were posted within minutes of the polls closing. Culberson was up by six points, 52 percent to 46 percent, with the balance to the libertarian candidate. It was clear Skelly's polls had been wrong. The gap didn't close. "We immediately knew we were toast," he said.

Walking into the club, into the gathering of friends and supporters, his faith was shaken. "In his mind, he thought people were going to vote for the person who was going to do something," his wife said, reflecting back on the election. "He truly believed that he was the best candidate." But the people of Houston had disagreed. It wasn't even close. The final vote was 55.9 percent to 42.5 percent; the libertarian took 1.7 percent. Skelly had taken an R+16 district and closed the gap by less than three percentage points.

It doesn't take a political scientist to see that any Democrat would've struggled in such a heavily Republican district. It would have taken years of laying the groundwork and getting to know local Democratic clubs. No one had done this work before Skelly. He was well-financed and smart, and had political charisma. But the early-twenty-first-century American political system, and Texas particularly, was gerrymandered to an extraordinary degree. Skelly could've been the second coming of John F. Kennedy and he still would've been a long shot. But Republican redistricting had made sure the Texas Seventh was out of reach to anyone but a Republican. It would take another decade of demographic change and political organization for a Democrat to unseat Culberson.

Many people who know Skelly believe losing was a blessing in disguise. They think he would have hated Congress. He wants to build things and do something that matters. Congress is where ideas go to die. Bills, reforms, policies, and bold ideas are anathema. Congress chews up idealists with endless compromise and committee meetings.

Skelly packed up campaign headquarters on Kirby. He had reams of unused "Skelly for Congress" letterhead and a growing desire to get back into energy development. He started thinking about the future.

Work on the expanded Katy Freeway finished and allowed Houston to continue sprawling westward. Acres of prairie were paved over to make way for subdivisions and strip malls. When Hurricane Harvey hit the region in 2017, there was less prairie and long-stemmed grasses to absorb the record rainfall. The runoff flowed eastward, engorging Houston's bayous and overwhelming two reservoirs. The Army Corps of Engineers released water from these reservoirs to prevent a catastrophic failure. Thousands of homes flooded.

9

Clean Line Energy Partners

M ichael Skelly was forty-seven, in good health and energetic. He and Anne had three children who were all in middle and high school. For the first time since a meeting with Donald Perry in Cambridge sent him on a road to Costa Rica and then into the wind business, Skelly didn't have a purpose.

He angled for a job in Barack Obama's incoming administration, hoping to be the assistant secretary in charge of renewable energy. He saw himself as an idealistic bureaucrat with a $1 billion budget. Why not? He had built about twenty wind farms at Horizon that produced more than five million megawatt hours of power a year. To put that into perspective, it was more power than that generated by a handful of smaller nuclear power plants.

Washington did not welcome Skelly with open arms. It was, he later said, a "truly humbling experience." He kept a spreadsheet to track his calls to Democratic bigwigs. Many went unanswered. When the phone calls didn't yield the desired result, he flew to Washington, where the incoming president's transition office was vetting applicants. He hung out at a nearby Starbucks, hoping, unsuccessfully, to run into staffers on a break.

Back in Houston, he took some meetings about developing more wind. People were willing to back him, but the prospect left him unenthusiastic. He wanted to move forward. He was starting to think and talk about climate change. He was plagued by the thought that all his work building wind farms wouldn't make a difference for the climate. "How are we ever going to move the needle on renewable energy?" Skelly wondered. At the time, solar was expensive and a niche product, an accessory for the wealthy and environmentally minded. It provided one-fiftieth of one percent of the electricity in the United States in 2009. Wind, a bit more mainstream, generated a bit less than 2 percent of the electricity in the United States. Skelly knew increasing that figure would be difficult.

A few months earlier, the U.S. Energy Department issued a report on renewable energy. It was titled *20% Wind Energy by 2030*. The agency concluded this was doable, but would require $20 billion for several thousand miles of new wires. The government report produced a map of where new transmission lines would be needed. In bright red, the largest cluster of new lines was centered in the thin sliver of the Oklahoma panhandle, heading eastward into Mississippi and Arkansas. The report stated: "If the considerable wind resources of the United States are to be utilized, a significant amount of new transmission will be required." Who would pay for it? The report was silent on this question.

———

Every generation or so, the country goes on an infrastructure-building spree to accommodate new forms of energy. New long-distance transmission lines owned by the federal government accompanied the dam-building era that bookended World War II. When giant nuclear and coal plants were built in the 1960s and 1970s, transmission lines were unspooled across the land to move power. The 1990s saw the rise of the superefficient natural gas plants, often built close to cities and industrial centers where existing wires were sufficient. But new pipelines were built to feed the plants with combustible gases.

Maybe it was wind's turn. Suedeen G. Kelly, a member of the Federal Energy Regulatory Commission, thought so. "We need an interstate transmission superhighway system," she told a reporter a month after the Energy Department report. The quote was in an article that described a wind farm called Maple Ridge in upstate New York that sometimes had to shut down even when there was a brisk wind blowing because of congestion on the grid. Skelly knew all about that wind farm. Horizon had built it.

The idea of a transmission superhighway system was the kind of grand vision that appealed to Skelly. It was the kind of thing that America once excelled at. The U.S. government and American companies had built the interstate highway system, the Hoover Dam, and the Panama Canal. Skelly loved these projects. He read books about them. The more he thought about it, the more he came to believe that the modern equivalent was a new and improved grid. It was certainly a needle mover, Skelly thought.

Around the time that Skelly was thinking this, unbeknownst to him and the public the federal government came remarkably close to embracing this idea and building a new superhighway for electrical power. On December 16, 2008, President-elect Obama held a private meeting in snow-choked Chicago to discuss the grim and precarious state of the economy. Christina Romer, an economist at the University of California at Berkeley, had prepared a PowerPoint presentation that made the case for a $1 trillion stimulus. She started off with a message designed to catch the president-elect off guard and make sure he paid attention. "This is your holy shit moment," she said. She and other economic advisors urged him to provide a defibrillating shock to the economic system to get companies hiring and consumers buying. Obama said a stimulus package couldn't just be about money, it also needed something to inspire Americans.

After a brief break for lunch, talk turned to energy. Obama repeated that the stimulus package needed something bold. "We need more moon shot," he said. He talked about how electricity from the

Dakotas, generated from the ample wind, could be brought into Chicago on high-voltage transmission lines. Vice President–elect Joe Biden said he liked that idea.

Carol Browner, Obama's top advisor on energy and the environment and a former head of the Environmental Protection Agency, dissented. Acquiring the land for these power lines would be difficult, if not impossible, and politically contentious. And there was the problem of the in-between states. The Dakotas would get wind farm jobs and lease payments. Chicago residents would get access to a new source of electricity—one that wasn't particularly expensive or subject to volatile price swings. What would Iowa and Minnesota get other than big transmission lines, she asked. The group discussed the idea of giving the states money to build parks.

There was one final argument countering Obama's idea to build a new power grid. To be effective, the economists reminded Obama that the stimulus needed to move money out the door quickly to begin circulating in the economy. A new power grid would take too long to plan. It was too susceptible to unexpected delays. By the time the meeting in Chicago was over, Obama's advisors had talked him out of the grid as a big idea.

———————

In the spring of 2009, Skelly was in a funk. He considered writing a book about the climate and energy. "The book was going to be entirely depressing and a complete bummer," he recalled, but thinking about it helped focus him. "People think that inspiration comes from, like, you're sitting around and you're happy about something and this great idea pops into your head. I think the opposite is true." Wallowing and worry were his sources of inspiration. "From despondency comes inspiration, not from giddy happiness," he said.

From Skelly's despondency came his desire to build transmission lines. They would begin in the Great Plains, that large airshaft in the middle of the country where the wind blew consistently and the wind speeds were strong. The lines would carry the renewable energy to where people lived. Conceptually, it was a simple idea. There was one large problem: He wasn't sure that a private start-up could build an interstate transmission line. Giant utility companies typically built them. They had authority to condemn property, if needed, to route these lines. And the companies didn't risk their money. The state utility commissions guaranteed them a return on the billions of dollars they invested.

Could a private company accomplish such an undertaking? Maybe. It was hard to say. No one had really ever tried. Out in Wyoming, billionaire Philip Anschutz's company was toying with the same idea, but hadn't progressed very far. Skelly figured someone had to test the idea to see if a private company could build a big interstate transmission line. Why shouldn't it be him? He figured someone would give him money to try because of his success at Horizon. And he would be building something challenging again. Skelly thought it sounded like fun.

If he could build a large transmission line to connect the windy and sunny parts of the country to the cities, renewable energy would take off. And if he could make money doing it, this would beget more investments and more renewable energy. If he could figure out how to do it, others would follow.

Texas had done something similar a few years earlier with remarkable results. In 1999, the state passed a bill that required utilities to buy renewable energy. Governor George W. Bush was a big supporter, encouraged by his friend and political booster Enron chairman Ken Lay. Most of the early wind farms were built around McCamey, a dusty crossroads in West Texas that had once boomed with oil prospectors who came, drilled, and then left. By 2001, there were six wind farms in the vicinity of McCamey and more on the way. A Republican state senator proposed a resolution proclaiming the town to be the "Wind Energy Capital of Texas."

Deb Lewis realized there was a big problem brewing. An ebullient woman with a blond mane and a loud laugh, she had earned a degree in electrical engineering from the University of Texas in 1983, one of twenty-one women in her class of 257 with the same major. About the time Texas lawmakers were singing the praises of McCamey wind, a client hired Lewis to look into whether to build a wind farm in the area. She advised against it.

The McCamey-area wind farms could generate 758 megawatts, but there was only enough room on existing wires to get 400 megawatts out of West Texas. The state's grid operator had allowed too many wind farms. It was like a county approving a new 2,000-home subdivision, but not considering that maybe it needed to widen the two-lane country road that connected the new neighborhood to the nearest major artery. Lewis arranged a meeting with the grid operator and said, "you guys are screwed and there is no stopping it." By 2003, the situation in McCamey was a mess and the Texas wind boom was teetering. Transmission equipment routinely got overheated and wind farms had to be curtailed. This undermined wind farm business models and enraged customers who had contracts for a certain level of power generation.

Texas decided to try a new approach. Instead of waiting for companies to build wind and then working on transmission, the state flipped the process. It would figure out the best spots in the state for wind development and build transmission to those areas before any wind farms. In 2005, the state legislature passed a law to begin work on what came to be called CREZ—Competitive Renewable Energy Zones. The idea was to build new power lines from West Texas and the Texas panhandle to allow inexpensive wind power to flow into the middle of the state, where the majority of Texans lived.

Originally estimated to cost $4.9 billion, the project grew to $6.9 billion. Texas ended up building 3,589 miles of power lines, more than enough to encircle the state. The lines spurred wind development. Thanks to CREZ, Texas leapt ahead of California as the top state for renewable energy. When the wind industry came to Houston in 2008 for its annual convention, Governor Rick Perry welcomed them. "You are in what we would call friendly territory," he said. "People who talk about wind energy as a technology of the future clearly haven't been to West Texas lately." By the end of 2008, the state had installed 9,000 megawatts of wind power; this would double within another decade.

One of the CREZ lines reached McCamey, allowing its power to flow north to the Permian oilfields and southeast toward San Antonio. These lines didn't end up just enabling wind development in the former oil town. Southwest of McCamey, the road goes over a slight rise and descends into a broad valley. There are wind turbines on the mesa and thousands of solar panels on the valley floor. Few Pecos cantaloupes are still grown in Pecos County. The new crop is renewable power.

———

There was nothing special about Texas's wind. The wind actually got a little better as you headed north from the Lone Star State. That was the thought running around Skelly's head. There was an inexhaustible supply of inexpensive renewable energy, and potential demand for it in the southeastern United States. Why not create a company to build the infrastructure to connect the two?

As his idea took root, Skelly had questions. What was the best transmission technology to use? Texas had relied on fairly standard 345-kilovolt alternating current for its CREZ lines, because they matched the rest of its power backbone. But Skelly reached for the best available technology. In late March 2009, he called ABB. The century-old Swiss company is known for is its high-voltage direct current power lines. There are two situations where DC lines make more sense than the more common AC lines. DC lines can go underwater; after a few miles underwater, AC lines become unreliable. And over long distances, DC lines lose less power. So if you want to move bulk power over a few hundred miles, direct current is the way to go.

When Skelly called, ABB was building a massive project in China, a 1,230-mile high-voltage direct current line that would carry 6,400 megawatts. That is enough power to light up the Las Vegas Strip, keep the tourists in the casinos at a cool 65 degrees on the hottest day of the year, and have enough left over for the rest of the city of 1.4 million people—on a single pair of lines. Skelly asked if the HVDC line could carry 7,000 megawatts, an early sizing of the project, for seven hundred miles. The answer was an unequivocal yes.

Work on the new company began in what Skelly called the "Casita," the little house. It was a romantic description of the garage apartment behind his home in the West University neighborhood. Skelly wanted to build something enormous—a power line that stretched across three states, held aloft by 150-foot-tall towers. But the small apartment was large enough to work on pitch decks and financial models.

Skelly's first recruit, Jimmy Glotfelty, required little convincing. At a family-run restaurant near his house, Skelly explained his idea over a Tex-Mex breakfast. "Wow," Glotfelty responded. "I've always wanted to start a transmission company." He had tried something similar, but less ambitious, a few years earlier. He lacked Skelly's entrepreneurial chops. "I was a government guy and I had a really hard time figuring out the business side of things," he admitted.

A native of San Antonio, Glotfelty joined the administration of Governor George W. Bush in the mid-1990s when he was a few years out of Texas Christian University in Fort Worth. He was assigned to work on deregulating the state's electrical market. When Bush went to Washington as the forty-third president, Glotfelty followed and took an appointment at the Department of Energy.

Over the years, Glotfelty had developed a good idea of how the power grid worked and didn't work. He was on vacation in the Santa Fe National Forest in 2003 when a blackout rippled across the Northeast United States, leaving 50 million people in the dark. He worked for days from a phone booth, the only working phone for miles, to coordinate the government response and investigate what went wrong.

This experience had opened his eyes. The Northeast United States had been dark, but there was ample power to the South and West. It didn't take an electrical engineer to see how a better-networked grid could have helped. After the blackout, he had participated in a brainstorming session at the Energy Department about a "National Electricity Backbone": building large transmission lines that would move power around from region to region. The idea sparked his imagination.

At the breakfast, Glotfelty explained to Skelly that at the DOE he had championed a way to attract private sector money to invest in this backbone. After he left Washington, the idea had made it into law as Section 1222 of the Energy Policy Act of 2005.

The federal government could partner with a private company to build interstate transmission. It was a way to modernize the grid without spending too much taxpayer money. The government would identify areas in need of transmission investment. The private company brought the money, and the federal government brought the power of eminent domain.

By the time breakfast was over, Glotfelty was ready to leave his job as a consultant and sign up with the new company. It could be a business that would make the U.S. power grid better, cleaner, more affordable, and more reliable. Glotflety started driving over to the Casita on Sunday afternoons to figure out how it would work.

Skelly was going to be the chief executive and visionary-in-chief. Glotfelty would figure out how to negotiate the byzantine state and federal regulatory pathways. What was needed was an actual developer, someone to focus on project minutiae. A month after his breakfast with Jimmy Glotfelty, Skelly had lunch with Mario Hurtado, an Ivy League–educated energy developer who had built hydropower dams in Latin America and consulted on liquefied natural gas terminals. As they ate at a Thai restaurant a little north of Skelly's house,

Hurtado asked Skelly what he was doing. Skelly started talking about the project. Hurtado said he was interested. He had spent nearly two decades inside corporations and wanted to work somewhere smaller and more entrepreneurial.

He liked Skelly's ideas. "Michael really had the vision of where wind was going to go and why transmission was the bridge to get there," he said. Before Hurtado and Glotfelty left their regular paychecks, Skelly convened a meeting of the three men and their wives. Anne Whitlock, Skelly's wife, says it was important for everyone to know what they were getting into before they got too swept up in the excitement.

Skelly was pitching and, she says, when he starts pitching he has a way of getting people to sign on to his vision. "Even if you have doubts, you think, maybe we can pull it off," she said. But she wanted everyone to have open eyes.

"It is a start-up and it is very risky. Our funding stream could disappear," she said. Anne had watched Skelly help build up Zilkha Renewable. There was luck and hard work involved. Nothing was guaranteed, and it would take several years before the project had any hope of being built. And the forces arrayed against the project could be considerable. No one had second thoughts. Glotfelty and Hurtado both decided to give it a shot and together with Skelly, they built the new enterprise.

After a few weeks at the garage apartment, Michael Zilkha offered Skelly the use of the conference room in his downtown office. Moving there was important, Skelly explained, because it was a much better place to raise money. "The best way to raise money is to look like you don't need it," he said.

Eventually, Skelly pitched the Zilkhas—father and son—on becoming the first investors in what the three cofounders called Clean Line Energy Partners. Skelly talked for twenty minutes about his vision for the company before Selim Zilkha interrupted him. He said he appreciated all that Skelly had done at Zilkha Renewable to make the company successful. But he was not interested. The financier-developer team that had built Zilkha Renewable into the billion-dollar Horizon Wind Energy was not going to partner again.

It would be too hard, said Selim Zilkha. For the right project, he would be happy to back him again. "This is folly. It's not the money, it's just I don't want to encourage you," he said. Skelly wasn't daunted by the warning. He was in thrall to a big idea that had gotten him out of his funk. There would be other investors.

Even after passing on backing the company, the Zilkhas told Skelly he was welcome to continue using the office space. To keep things going, Skelly began to bankroll the project. He sank hundreds of thousands of dollars into the effort, paying for studies and interconnection requests. He eventually funded payroll.

Clean Line still didn't have a permanent office space or a long-term funding source, but it soon got a summer intern named Charlie Ary. An evolutionary biology major at Rice University, he was supposed to work at Skelly's old company, Horizon. But a friend of Skelly's thought he would find it more rewarding to work at Clean Line. Ary was traded to Clean Line like a minor league baseball player. He ended up working there for seven years. His first duty was to create a filing system. He kept records, printouts of correspondence, and iterations of financial models. To save money, he used leftover campaign stationery. Early Clean Line documents often had a purple "Skelly for Congress" logo on the back.

Charlie Ary wasn't the first in his family to be in the energy business, although he kept this to himself. His grandfather was T S Ary, spelled without periods. An unabashed supporter of mining, he rose to become head of the U.S. Bureau of Mines under Presidents Ronald Reagan and George H. W. Bush. He once swallowed a nugget of galena, an ore use to make lead-based paints, to prove to officials that it was inert and didn't need new regulations governing its shipment. Another time, in 1991, in a speech to farmers, miners, and oilmen in Denver, he said he didn't "believe in endangered species."

Charlie Ary says he doesn't know much about his grandfather, whom he called Grand T. Like many of his generation, Charlie was driven by a strong belief that renewable energy was the future. He was the first of many idealistic young people who came to work at Clean Line. Skelly knew this generational preference for renewable energy was a competitive advantage for the company. He could attract young energetic talent in a way that a coal company couldn't.

———————

Skelly, Hurtado, and Glotfelty began pitching investors in July 2009. Skelly shared his idea with Goldman Sachs as well as other investment banks and private equity companies. He talked about the transmission lines they planned to build as "renewable energy pipelines"—using the language of oil and gas that would be familiar to investors—that would connect "quality renewable resources and energy demand centers." The first line, originally called Great Plains, "taps into area rich in wind and solar resources." Developing and permitting the line would take $50 million, but they only sought half of that to get moving. Once approved, it would cost $3.5 billion to build.

Clean Line planned to develop five different lines and begin construction on the first in 2015. The founders decided to pursue multiple lines so that if one encountered an immovable political obstacle or an entrenched utility determined to sabotage the project, it was expendable. "Even one successful project will justify mid-term investment," the company's pitch deck concluded.

The first line would show it was possible, encourage financial backers, and put regulators' minds at ease. After that, more lines could be built until Skelly, Hurtado, and Glotflety's vision of a new electricity backbone with many high-voltage power lines became real. It was an ambitious goal, and it created multiple projects that required attention and capital. Skelly later said that trying to build fewer lines at the same time could have saved them headaches in the future.

Clean Line's plan wasn't particularly technically challenging. They weren't exactly going to Home Depot and buying all of the available extension cords, but it was more or less off-the-shelf technology from industrial giants like ABB and General Electric. The problem, Skelly figured, would come from trying to change a sclerotic electricity system hostile to outsiders and new ideas. U.S. electrical grids didn't have a place for entrepreneurs like Skelly.

Skelly figured that utilities would resist the new transmission lines. People who work for utilities tend to be conservative by nature. "They are not the freelancers of the energy world, right? They are not the wildcatters. They are the folks who want a steady job and often a fixed-benefit retirement," he said.

They are not rewarded for taking risks, he said. "They work within a regulatory environment that doesn't reward them for extraordinary performance but punishes them for mistakes," he said. "And they are not exposed to competition. So therefore they don't have a need to embrace new ideas."

———————

Skelly practiced his sales pitch on Ary. Clean Line would open up the potential for more wind and solar development, Skelly said in these dry runs. This was both an opportunity to make a good return and an opportunity to do something good for the country and the environment. Skelly had already turned wind farms into a small fortune. Now he could repeat it with renewable energy infrastructure. Ary was impressed by the breadth of the vision.

Most of Skelly's pitches didn't end well. Investors were afraid because no one had built anything like this before. Some meetings went very poorly. In San Francisco, Skelly met with representatives from the Texas Pacific Group, a large private equity firm. As Skelly walked them through the pitch, the prospective investors ate catered sandwiches. Over thirty minutes, Skelly explained how the price difference between the cost of generating wind and the price in eastern markets was high enough to more than cover the cost of long-haul transmission. As he finished the pitch, he sat down at the long conference table. The investors thanked him and excused themselves. "They left me to finish my lunch by myself," Skelly said.

At another meeting at the Four Seasons Hotel in downtown Houston with representatives of the Ziff family, Ary handed Skelly the presentation. As Skelly was passing it to Ziff representatives Bryan Begley and Neil Wallack, Ary saw a flash of purple. "My heart sank," he remembered. Each page of the investor slides had "Skelly for Congress" printed on the back. Begley had a good sense of humor about it. He leaned over to Ary. "This is your first business meeting, isn't it?" he asked. Ary admitted it was. Then Skelly jumped in. We're a lean start-up. We got a free intern and we recycle old paper. We're not going to waste the investors' money, he told them.

Despite the awkward start, the Four Seasons meeting brought Skelly together with Clean Line's first financial backers. Begley, literally an Okie from Muskogee who had gone on to Harvard Business School and McKinsey and Company, scouted energy investments for the Ziff family, which had made a fortune in magazine publishing. In November 2009, Ziff's investment vehicles put $25 million into Clean Line. By this time, Skelly estimated he had put in $1 million of his own money.

There was a lot that Skelly couldn't put in the pitch deck. Like how he believed life was at its richest when you did things that were big and bold and challenging. "If you are motivated by that stuff, it doesn't have to work for it to be worth it," he said. "If you're an investor, that's probably not a great answer. . . . But what else are you going to do other than the stuff you believe in that is arguably important?"

By the end of the year, Clean Line was a real company. It had money and a plan. It even had an office, in the same building but a few floors below the Zilkhas'. Skelly hired developers, a corporate counsel, and people who understood high-voltage direct current technology. Appropriately enough, given their mission, a clutch of electrical wires hung from the unfinished drop ceiling in their conference room when the company moved in.

Some of the new workers pitched in to buy an office ping-pong table. They set it up in the kitchen, and used it to blow off steam. When one of the developers, who could put a serious spin on his serve, played against the new chief financial officer, who had a ruthless forehand smash, work around the office stopped to watch the match.

A branding company was hired to brainstorm names for the different transmission lines. They suggested names reminiscent of railroads. "In the same sense that the railroads connected the country for transportation, we were going to do the same thing for electricity," recalled Ary. Great Plains became the Plains & Eastern Clean Line. Another line that would start near Dodge City, Kansas, near the Oklahoma panhandle, and run to the Illinois-Indiana border, became the Grain Belt Express Clean Line. A third line that would cross Iowa and send power into Illinois was the Rock Island Clean Line. A fourth would head west across New Mexico and was named the Western Spirit Clean Line.

Clean Line started holding meetings in Oklahoma and Arkansas and later Kansas, Missouri, Iowa, Illinois, and New Mexico. They decided to give landowners as much information as possible to calm fears and, hopefully, win over residents. It would soon become clear there was a large downside to this approach. Clean Line was basically giving opponents lots of information to fight the project if they wanted. And because they started community meetings early in the application process—instead of waiting as long as possible—they were giving communities more time to organize. This decision would come back to haunt Clean Line in years to come.

Skelly also decided that every employee had to attend community meetings on behalf of Clean Line. It wasn't just going to be Skelly and Hurtado. The lawyers went and so did engineers. Some employees attended dozens. Getting yelled at by local landowners wasn't fun, but it instilled a sense of mission. It wasn't long before Ary hit the road with Clean Line. A month after graduating from college and becoming a full-time employee, he woke up early to make sure the Mendota Civic Center, in north-central Illinois, was open at 7 a.m. with maps spread across several tables. And he stood behind the maps, explaining to whoever showed up why the project was so important. And he manned the booth at the Clay County Fair in Iowa. "Everybody loves wind energy here," he enthused. But this wasn't always true. Not long before, he had been yelled at while attending a different open house. He found the disconnect between his idealism and the anger he sometimes faced difficult to process. But that was part of the job.

10

Euphoria

Michael Skelly embraced a theory of project development that he liked to share at annual holiday work parties and on other occasions when the situation called for it. The long process of taking an idea and turning it into a fully permitted project had five distinct stages. The first stage was euphoria. You come up with an idea and everything seems to click into place like tumblers inside a lock.

In 2009, Michael Skelly was in this stage. He had an exciting idea. There would be hidden shoals lurking somewhere ahead, unanticipated setbacks that could detour the best plans. But those were future problems that belonged to the second, third, and fourth stages of projects: despair, the search for the guilty, and the punishment of the innocent. The fifth stage was riches and glory for the uninvolved. The stages would get more arduous. To persevere, Skelly would need a good team who were devoted to the idea of tapping the wind.

He wanted to hire smart, committed employees. He wanted people he believed in and who believed in the mission. He wasn't promising the best salary. And the job would require hitting the road in Middle America and getting yelled at a lot by angry landowners. "Look, it is hard as shit," Skelly told people he wanted to hire. "But it is worthwhile because we are tackling one of the biggest challenges of the day. It is that simple." Some people were turned off, but most who heard the pitch wanted to sign up.

Several years earlier, the Zilkhas had recruited Jayshree Desai to join Skelly at Horizon. She was a corporate number cruncher. "I remember thinking that wind isn't going to go anywhere, it was a bunch of greenie weenies," she said. But she had met with Skelly and liked what he said about using the private sector to raise money and try to change the world. After Skelly left Horizon, she remained as chief financial officer with the new Portuguese owners.

"I had a serious job," she said. She liked her co-workers and the European business culture. Skelly wanted her to join Clean Line and put the hard sell on her. She struggled with the decision. A few years earlier, she had been at Enron, a young MBA involved in buying and selling energy assets. She didn't realize the company was lying to its shareholders until it was too late. The experience left her searching for a company interested in more than just making lots of money. She also worried that having Enron on her résumé would scare off future employers. Skelly and the Zilkhas had taken a chance on her then, setting her career back on track. Now he was asking her to take a chance on his latest venture. She agreed, "out of personal loyalty to Michael," she said. She would become chief operating officer of Clean Line. She became Skelly's sounding board—and unofficial therapist.

Skelly liked to hire corporate refugees: people who had grown frustrated working for a large and powerful employer. The kind of candidates Skelly pursued were willing to walk away from the large paychecks and annual bonuses in search of something more fulfilling.

One such refugee was an early hire named Cary Kottler. He was a lawyer who had spent his first few years out of law school on the partner track at the giant Houston energy firm Vinson & Elkins. One day he realized he didn't want to make partner. Kottler compiled a list of people in Houston working on renewable energy. There were two names on the list, and Skelly was at the top. Kottler called and set up a meeting at Starbucks. When he arrived, Jayshree Desai was there also. Skelly got a phone call and left. Kottler stayed and talked to Desai. He liked what he heard and took a job, not as a lawyer, but as a developer. Within a year, he went from working in the comfort of a fortieth-floor office in a Houston skyscraper to attending rural Rotary Club meetings. "I took a chance on him and he took a chance on me," Kottler said of Skelly.

If the new hires weren't corporate refugees, they were often young, recent graduates of some of the country's best colleges. They were committed to building renewable energy and fighting climate change, and turned down more money to come to Houston and work hard.

Diana Rivera joined soon after Desai and Kottler, in spring 2010. With an industrial engineering degree from Cornell, a couple of years working at General Electric, and an MBA from Harvard, she had a buffet of career choices. One day, at Harvard, she attended a meeting of alumni talking about energy, the environment, and business. She met Skelly after his panel. He looked up her résumé during the next panel, interviewed her in the hallway, and offered her a job on the spot.

After graduation, she joined Clean Line and turned a small round table into her desk. The cords were still hanging from the ceiling of the unfinished office. She didn't mind. She liked the work, the company's culture, and the challenge.

Clean Line would soon need to go before state regulators and apply to become a power utility. To be taken seriously, it had to become a serious company. Show up in Little Rock or Oklahoma City as a start-up with a handful of employees, and it would be giving skeptical regulators permission to turn its application down. The payroll expanded rapidly, and as Clean Line added employees, its operating costs grew. It began to burn through investors' money.

It was a start-up, but Skelly once sent an email around advising Clean Line employees not to refer to the company as a start-up. We should try to "punch above our weight," he said. They would encounter powerful utilities and politicians in coming years, he said, and as Clean Line employees we need to take ourselves seriously. If we don't, he wrote, no one else will.

As Skelly and the growing Clean Line team pursued the audacious goal of developing renewable energy projects measured not in the hundreds of megawatts, but in the thousands, timing appeared to be on their side. Interest in renewable energy was growing. World governments were preparing to meet in Copenhagen to discuss what to do about climate change. The International Energy Agency, the world's preeminent energy watchdog, declared the world at a "crossroads." Down the status quo path was more fossil fuel burning and possible "catastrophic and irreversible damage." The other path involved using energy more efficiently and rapidly switching to renewables.

This message buoyed Skelly. But it was one thing to write reports about the need to change to a low-carbon energy world. Skelly wanted to build an actual power line across entire states. That meant the new company needed to attend to a myriad of details. One of the most pressing was where to deliver the power. An interstate power line would need to connect to a substation—a large electrical switching yard where huge amounts of power were divvied up into smaller pieces for distribution, like a butcher turning a carcass into wax-paper packages of beef ribs, steak, and brisket. But which substation?

Skelly paid out of pocket for consultant studies to suggest the best place for his company's proposed power line to end. One possibility was in Arkansas. Another was to cross the Mississippi River and deposit the power near Memphis. Skelly and Mario Hurtado flew into Memphis. They circumnavigated the city in a rental car, searching out substations that were sometimes designed to be hard to see in order to keep as low a profile as possible. They then headed west on Interstate 40 toward Little Rock. There they caught up with a large 500-kilovolt line headed south to Louisiana and toured more substations, stopping in at El Dorado, Mabelvale, and Sheridan like a touring music group willing to play as many small venues as they could.

They spent three days in the car, eating barbeque and looking skyward to follow the power lines until they came earthward. Finding them could be tough. "It was like a treasure hunt," said Hurtado. Visiting the substations was important; otherwise they were just circles on a topographical map. "You need to go and physically look at stuff. You never know what you are going to see," he said.

In the end, they decided Tennessee was the best option. It was the shortest route to get the power into the southeastern United States. From Memphis power could move in several directions to the big cities to the east and north. The Shelby substation on Mudville Road, north of Memphis, seemed the most promising. It was twenty acres of transformers, wires, switches, and breakers amidst open fields. The line would stretch more than seven hundred miles from the Oklahoma panhandle, carried by slender monopoles until it came to ground at the edge of the power grid operated by the Tennessee Valley Authority. The timing again appeared good. TVA had just experienced a traumatic incident that had shaken its confidence in coal, a fuel it had relied on for many years.

———

When the Kingston Fossil Plant opened in 1954, it was the world's largest coal-burning power plant. It took in and burned 14,000 tons of coal daily. TVA trumpeted that the plant turned this prehistoric material into 1,600 megawatts of electricity, enough to power Memphis. It was quieter about another daily output: 1,000 tons of coal ash, a toxic by-product of burning coal.

To handle the stream of watery waste, the TVA built a settling pond to collect the coal ash. About a decade after the plant opened, the pond filled up. Another pond was built and then another, impounded by a series of dikes and cells. An industrial ecosystem emerged: coal went into the plant and was burned and ash came out. The waste grew into a geological feature on maps.

As more ash slurry entered the pond, particles settled to the bottom. Water was skimmed from the top. In this way, slowly, over the years, a hill of ash grew, with a pond on top. The TVA raised the dikes to accommodate the rising level of waste. In the 1950s, the pond's elevation was 748 feet above sea level. By the 1970s, the utility raised the elevation to 765 feet and then 774 feet. The growing weight of this pile of ash began to create structural stress. By the 1990s, seepage was a recurring problem, solved by a series of clever engineering fixes involving riprap, interceptor trenches, and, finally, large drains. By this time, the ash mound had risen to 866 feet above sea level. There were plans to go even higher.

The cycle of coal trains disgorging their cargo, machines pulverizing and burning tons of Appalachian fuel, turbines generating electricity, and waste collecting in ponds reaching skyward continued for more than five decades. Shortly after 1 a.m. on a bitter December 22, 2008, the cycle ended catastrophically.

The accumulated weight of fifty years of coal detritus slid sideways, destroying a set of cells and dikes, and sending a forty-seven-foot-tall wave of coal ash smashing toward the Tennessee River. It was the worst industrial spill in the history of the United States, a record that would stand for only sixteen months until the Deepwater Horizon oil spill in the Gulf of Mexico. A four-volume analysis into the root cause of the TVA disaster would later point out that the coal mountain was built atop a thin base of slime less than six inches thick. It was as if the TVA had built a multistory office building atop a layer of grease.

About 5.4 million cubic yards of coal ash spilled out, a witch's brew of metals, toxins, and potential carcinogens, including antimony, arsenic, beryllium, cadmium, chromium, copper, lead, mercury, nickel, silver, selenium, thallium, vanadium oxide, and zinc. Three homes were rendered uninhabitable and another twenty-three damaged. There were no injuries reported, at least not initially. Later, workers employed in the cleanup would sue for long-term damage to their health.

By the time the sun rose on December 22, a gray moonscape came into view where a lush river bottom had existed a day earlier. Cleaning up that mess would cost $1.2 billion. Coal had always been viewed as an inexpensive source of power at the TVA. The cleanup raised serious questions about this calculation. Coal was inexpensive to buy and burn, but there was an unpaid, overdue bill that had come due.

Another kind of invoice was dangerously adding up. In the spring of 2009, a group of German, Swiss, and British scientists published a peer-reviewed academic paper with a stunningly simple idea that tugged at the fuel-burning fabric of modern society. If the world wanted a reasonable chance of limiting rising temperatures to no more than 2 degrees Celsius, the authors asserted, humans had to go on a crash fossil fuel diet. Two degrees was a climate Rubicon. Cross it and consequences become irreversible. But humans had already found enough new supplies of coal, oil, and gas that if it all were burned, the temperature would smash through 2 degrees. In other words, the cycle of finding, extracting, and burning fossil fuels had to be broken. No more than a quarter of known reserves could be burned; most had to remain in the ground, the authors argued. Other sources of energy that didn't emit carbon dioxide had to be developed by industry. Government also had an important part to play. New policies were "needed urgently" if the 2-degree target would be met, the paper stated.

A couple of weeks after this paper was published—and greeted by widespread news coverage—Skelly and Jimmy Glotfelty flew to Washington, D.C. Glotfelty had secured a meeting with Steven Porter, a lawyer at the Department of Energy. They were delivering a heads-up to the agency. Clean Line planned to build at least one, and most likely several, new interstate transmission lines. And they would be seeking to partner with the federal government and ask it to use its Section 1222 authority. All that Skelly and Glotfelty were asking for at the time was for the Energy Department to get ready to consider the project.

He expected a positive reception. The new president, Barack Obama, had used his first weekly radio address to talk about the need to "accelerate the creation of a clean energy economy" by doubling wind and solar power. He talked about the need for new transmission lines to carry this energy from "coast to coast." The federal government's interest fed Skelly's euphoria.

Skelly and Glotfelty arrived at the Department of Energy's hulking headquarters in Washington with what he thought was good news. The private sector was ready to step up and spend money to build the new transmission that President Obama had talked about. No taxpayer money was needed, he said, and electricity rates wouldn't have to be raised to pay for the projects.

As far as Skelly remembers, the meeting went well. Porter took down the information and asked good questions. There were some legal issues that needed to be considered, but the first step was for the Energy Department to issue an announcement to see if any other company was considering something similar. Skelly left the meeting feeling confident that this bureaucratic step would happen soon.

A few days after Skelly's Washington meeting, Skelly asked Clean Line's indefatigable intern, Charlie Ary, to fax over an interconnection request to TVA. Skelly sent over a personal check for $10,000. We want permission to add electricity to your grid, the request said. Ary had never used a fax machine and struggled to get it to print out a receipt so Clean Line had proof of the request.

The request was to connect to the TVA's Shelby substation. A week later, Clean Line provided the TVA with more details about their plans. The wind farm we're talking about is "without precedent in scale and ambition," it wrote. We'll sell you as much renewable energy as you want, the letter said.

The TVA is the largest public power system in the country, providing power to 7.5 million people in Tennessee and parts of six other states. More than half of its power came from burning coal. After the Kingston mess, the TVA began to take a hard look at renewable energy. How much would it cost to generate a megawatt hour of power from the wind or the sun? How could the TVA system operators incorporate it onto their power grid?

The TVA—and the rest of the southeastern United States—had never relied on wind or solar power. If you look at a map of wind farms in the United States, there are clusters of dots throughout the Midwest, Great Plains, and California, but fewer than five isolated dots south of Charleston, West Virginia, and east of Dallas, Texas.

The wind just doesn't blow reliably in the Southeast. The one wind farm in TVA territory proved this. Erected on Buffalo Mountain outside Knoxville, three small turbines operated from 2002 until 2009. But they only ran a dismal 18 percent of the time. A set of newer turbines, erected in 2006, did marginally better—only 20 percent of the time.

It was a poor place to build wind turbines. Years later, Michael Polsky, the head of Invenergy, the Chicago company that built the second batch of turbines on Buffalo Mountain, said the wind forecasts were too optimistic. The wind didn't blow much. The company spent $40 million on the project and lost $6 million, he said. There's a reason the nearest wind farm to Knoxville was 250 miles away, in West Virginia. The southeastern United States doesn't have good wind.

The only place where there is even modest wind was atop mountains and ridges. But that meant the turbines would be highly visible. When TVA looked at wind a couple of years earlier, it concluded covering mountains with wind turbines would have a major visual impact.

Even before the Kingston spill, the TVA was taking another look at wind, but with a new twist. Instead of putting turbines up on the region's mountains, they planned to contract with wind farms hundreds of miles to the northwest. In the months after the spill, the TVA looked into long-term contracts to buy wind harvested in Illinois, Kansas, and Iowa. The price was steep—about $80 to $85 per megawatt hour. But TVA didn't think it had much choice. Its coal plants were under pressure to clean up their emissions that dirtied the air with sulfur oxide and nitrogen oxide, an expensive endeavor.

Skelly warned the TVA in May the wind contracts it was

considering were naive. There wasn't enough transmission to get the power from the Midwest back to Tennessee. TVA was buying nonfirm transmission, which meant they could use the lines, but would get bumped when the lines were otherwise occupied. As more wind is built in Iowa and elsewhere, the lines would clog up, Skelly warned. The TVA would have to pay congestion charges to move their power— or end up not being able to utilize the power at all.

"Without a dedicated transmission, fixed cost transmission path . . . Clean Line believes that TVA should carefully consider the likelihood that this energy's value will degrade as more and more wind projects are built in these areas," Clean Line wrote to the TVA.

Skelly urged the agency to take a hard look at his high-voltage direct current line. Rather than generate power in Kansas and rely on a bunch of small roads, some with tolls, to get the electrons into the Tennessee Valley, the TVA should use the superhighway that Clean Line was building.

It would be an "express train" from the best wind in the country straight into the TVA grid, Clean Line wrote to the TVA. It was quoting a large study issued a year earlier that TVA had participated in. If you build too much wind capacity without enough transmission, then local prices in windy regions would crater, undermining the power markets and creating havoc. But build a high-voltage direct current line, or HVDC, and the system could balance reliability and economics. And that's what Skelly was offering. There wouldn't been any depressed pricing or line congestion.

Skelly wrote a letter that accompanied its proposal with what he expected to be the big draw. "Clean Line expects to deliver renewable energy for a charge of approximately $25 per MWh [megawatt hour]. Based on our experience in developing wind projects, we expect production costs for wind from regions to be approximately $40 per MWh, for an all-in cost of $65 per MWh, delivered directly to TVA's substation." TVA was still negotiating its wind contracts for the Kansas, Illinois, and Iowa wind farms. Nothing had been signed yet, but

Skelly's wind was substantially less expensive than the $80 to $85 bids TVA had received.

Skelly flew to Tennessee and met with Rob Manning, TVA's executive vice president of power systems operations. A quiet-voiced North Carolinian, he had begun working at Duke Energy as an electrical engineer right out of college. The care and maintenance of power grids was his professional life. Skelly pressed his point: Our wind is cheaper.

Manning worried about Skelly's proposal. He was offering a large slug of electricity near Memphis. Would it destabilize the TVA system? "We were talking about 3,000 megawatts, potentially, dropping into a single spot on an electric grid," he said. "If the line tripped offline, how do I immediately replace those 3,000 megawatts?" Within a few months, Clean Line and TVA had signed a deal to jointly explore the proposal and evaluate any technical challenges.

That summer, Skelly went to Oklahoma to brief state officials there also. Oklahoma sits atop some of the richest oil and gas fields in the country. Near Tulsa, several major oil pipelines come together, and oil is stored there by the millions of barrels. It is such an important oil crossroad that the price of oil traded there sets the price of oil across the country, and to an extent across the world also. Skelly spoke the language of fossil fuels. What he proposed building, he said, were "renewable energy pipelines." Western Oklahoma, he said, was blessed with both a good wind resource as well as a strong solar resource. He also emphasized that what he was thinking about was oil-and-gas-sized investments: $3.5 billion to build the line and $10 billion in other investment.

Working through a PowerPoint presentation, he clicked to a map Clean Line had put together. At this point, they were calling the project the Great Plains Line. The name Plains & Eastern would come later. It showed the line originating in the Oklahoma panhandle and heading eastward, following the Arkansas-Kansas state line until it crossed into Arkansas and then headed east-by-southeast toward Memphis.

Wind power is changing, Skelly explained. The wind farms of the size he had started out building—Tierras Morenas in Costa Rica was

20 megawatts and Somerset in Pennsylvania was 30 megawatts—were gone. It was time to think big, he said. Wind farms that generated 1,500 to 10,000 megawatts "are now within the realm of possibility, reaching economies of scale required to support long-haul transmission."

He returned to Houston brimming with excitement. What had been a dream built from scraps of paper and a few ideas was starting to gain traction. It was around this time that Skelly met Jay Lobit, the Oklahoma wind developer who invited him out to Guymon.

He flew to Amarillo with Mario Hurtado. They drove north and met Lobit in a small town for a barbeque lunch. They headed north, crossing into Oklahoma, and let Lobit show off some of his early wind projects. Hurtado remembers the immensity of the flatness. "It was massive. You get out of Amarillo and you drive a good hour and a half before you hit a town," he said. "It is just obvious that wind is everywhere. I don't know if it was especially windy that day, but it was pretty windy." What he saw encouraged him, but also gave him pause. "It was kind of scary," he recalled. He had worked on some big projects before, power plants in Latin America and a liquefied natural gas plant on the Texas coast. But the enormity of the scale of what they wanted to do was sinking in. "I remember just looking at it on a map and going, 'Yeah, this start-up will build a three-state transmission project,'" he said. "It was clear that this was a huge climb."

Hurtado was edging into the second stage of project development: despair. Reality was setting in. Skelly remained in the euphoric stage. Things were looking up, so much so that Skelly and the rest of the growing team at Clean Line didn't pay much attention to a speech on the floor of the U.S. Senate by Lamar Alexander from Tennessee. A former college president and presidential aspirant, Alexander was reacting to a speech that President Obama had given a day earlier at a climate change summit at the United Nations Headquarters in New York. "After too many years of inaction and denial, there's finally widespread recognition of the urgency of the challenge before us," Obama had said.

Alexander said he was still shaking his head in disbelief. He noted that Obama had talked about generating increasing amounts of electricity from the wind. "These are the giant fifty-story wind turbines that they want to string along the Appalachian mountaintops from the Smoky Mountains and along the coastlines, which require transmission lines through your backyard. That's the plan," he said.

What Skelly and everyone else at Clean Line failed to appreciate is that the senior senator from Tennessee—a federal lawmaker from the same state they were relying on to buy their wind—really didn't like wind power. In fact, Alexander had an inalterable aversion to wind. But Clean Line didn't want to erect wind turbines on Tennessee's mountain ridges. So Lamar Alexander's stated opposition wasn't aimed at them, Skelly figured.

In October 2009, Clean Line signed a memorandum of understanding with the TVA to explore the technical challenges of connecting to the Oklahoma-to-Memphis line. The TVA was talking with them and the Energy Department was receptive. They had found the right place to gather up the wind.

————————

"Do you think we're crazy doing this?" Skelly blurted out at a meeting in the Atlanta airport in November 2009 with Terry Boston, an elder statesman in the electricity world. It was one of many meetings he arranged around this time to talk about Clean Line and its plan to build direct current power lines, to gain intelligence, and to look for allies. And he asked this question, or some variation. Sometimes he would ask: "Is this completely nuts?"

Boston's opinion mattered. He was a respected figure in the power industry. He had grown up on a farm in Tennessee, attended Tennessee Tech, and then went to work for the TVA. He had spent thirty-five years there, rising up from a project engineer writing software that remotely controlled power plants to executive vice president of the organization. Then he had moved to Pennsylvania to run the PJM Interconnection, a successor organization to the PNJ

Interconnection, the giant power grid that covers a large portion of the eastern United States. More than 20 percent of the economic activity in the United States uses electricity generated on the grid that Boston ran. There were few people more knowledgeable not just about the engineering of power lines, but the political power that governs them.

Boston paused and laughed. Thomas Edison was right, he believed. We need more high-voltage direct current lines. He had lost that fight with George Westinghouse a century earlier. But maybe the famous inventor would end up being proved right—again. Europe and Asia were bringing back long-distance direct current lines, why not the United States? Boston had even pushed to build giant direct current lines from the North to the South, which would have stopped the falling dominoes that caused the giant 2003 blackout.

Boston chuckled. He knew it would be really hard. He told Skelly he wasn't crazy. It makes sense, he said, but it wouldn't be easy.

Mario Hurtado remembers the exchange well. He was struck by how Skelly always asked this question, or some variation of it. It wasn't insecurity. He was probing, trying to get an honest reply.

"If thirteen out of fifteen people had said 'This is completely nuts. You shouldn't be doing this,' I think he would have thought about it. But that typically wasn't the answer we got," said Hurtado.

11

Chickens and Eggs

Within a year, Clean Line grew from a small group that could hold an all-staff meeting around a breakfast table in a garage apartment to thirty employees. With its payroll, Clean Line's ambitions grew.

The Plains & Eastern line was the first of several direct current lines Skelly and his colleagues planned to develop, each extending out from the windy plains. Maps of possible transmission routes were tacked onto the walls of the Clean Line offices alongside some of Michael Zilkha's collection of rock 'n' roll photographs. Skelly, Hurtado, and the rest of the company decided the Plains & Eastern would start near a substation south of Guymon, Oklahoma, and end at a substation north of Memphis. With a ruler, that was a 650-mile straight line. Without adding a parabolic, expensive detour, the route would go through Oklahoma and Arkansas.

Clean Line needed the states' permission to build and operate electrical equipment. This wasn't a simple task. Clean Line wasn't a power company in the eyes of either state. One of the first orders of business for Skelly and Jimmy Glotfelty was to convince regulators in

Little Rock and Oklahoma City that Clean Line was as trustworthy and necessary as any homegrown utility.

In May 2010, Clean Line filed an application with the Arkansas Public Service Commission to operate as a public utility. The document spells out Clean Line's objective and concept. The average wind speed across the entire southeastern United States—from Louisiana and Arkansas to Georgia—was poor. On average, it didn't go above 14 miles per hour. But further west, in Texas and western Oklahoma and southwestern Kansas, winds regularly topped 20 miles an hour. A wind turbine in eastern Arkansas would generate power *maybe* 30 percent of the time. In Oklahoma, the same turbine would spin nearly 50 percent of the time or more. Clean Line didn't say this in its application, but it was a similar story for solar. The good sunny locations were in Texas, Oklahoma, and New Mexico—the westernmost edge of the giant power grid that covered the eastern half of the United States.

"Our mission is to connect renewable energy resources to load centers in the most cost-effective way possible," Skelly said in testimony he presented to the three-member Arkansas Public Service Commission. A load center was a way of saying an area that uses a lot of power, such as a city or an industrial park.

Skelly was aware that regulators in Arkansas might not look favorably on the project. The wind turbines would be erected in Oklahoma; the electricity delivered in Tennessee. Arkansas would bridge the two. So he tried to turn what appeared to be a negative into a positive. "Geography places Arkansas between some of the windiest states in the country and large electric loads," he said. But with its ports on the Mississippi River and interstates, Arkansas was an "ideal" place to locate manufacturing plants to build parts for the wind industry. Two separate companies had already done so and could eventually add 2,500 jobs in the state, he told regulators. What's more, tapping domestic wind would improve the country's energy security. It would reduce pollution. And it would foster more power competition, bringing down electricity prices. But Clean Line wanted a right-of-way, cleared of trees, that would look like a giant lawnmower had cut a hundred-foot wide swath across the state's woodlands and rice farms. Regulators would need convincing.

Skelly wanted Arkansas to grant his upstart out-of-state company the same rights and responsibilities as the old Arkansas Power and Light Co., which had since changed its name to the more modern, solid, and corporate-sounding Entergy Arkansas. But long before the name change, Arkansas Power and Light had left a booby trap embedded in the law. And Clean Line had walked right into it.

Arkansas Power and Light had come into being in 1913 when a local entrepreneur named Harvey Couch purchased bankrupt power companies in the towns of Arkadelphia and Malvern. As was common at the time, most towns generated their own power, and these two towns southwest of Little Rock were no exception.

Couch decided to connect the systems, much like Thomas Edison's confidant, Samuel Insull, had done a few years earlier in Lake County outside Chicago. In Malvern, Couch built a power plant that used waste steam from a local lumber mill to generate power. He built a twenty-two-mile power line to carry the new power from Malvern to Arkadelphia. To inaugurate and advertise the new system in 1914, he hired two "Electric Queens" to travel between the towns, wearing a crown of light bulbs. In each town, the queens flipped a switch and there was round-the-clock electricity for the first time.

Couch's business strategy was simple and remarkably similar to Skelly's. He found inexpensive sources of power—waste steam from a lumber mill, for instance, or a seam of coal in Russellville, Arkansas—and then constructed transmission lines to spread the supply of power. Within a decade, the company operated four hundred miles of power lines.

By the mid-1930s, Couch's fast-growing company served cities in Arkansas and in neighboring Mississippi and Louisiana. In Little

Rock, the company ran advertisements in local papers urging people to keep the lights on. "Don't cheat yourself out of Better Living by skimping on electricity," the ad copy read. "It is your cheapest and best servant!" The company focused on the region's cities and towns, anywhere that had enough people to justify investment.

Outside of Arkansas's urban areas, the majority of farms did not have access to electricity. There were too few farms to justify the cost of constructing the necessary power lines. When private companies balked, the federal government created the Rural Electrification Administration to help. Couch distrusted the federal government and worried this new program was designed by bureaucrats to undercut and eventually take over his company. In a defensive move, Couch and his company defended their home turf in 1935 by backing a bill to create a Department of Public Utilities to make sure his utility would be regulated out of Little Rock, not Washington, D.C.

Couch and Arkansas Power began their own rural electrification campaign. He urged farmers to add more hens to pay for electricity. "More hens more eggs. More eggs more cash. More cash more kilowatt-hours," the company said. It was known, more simply, as the Eggs-to-Kilowatts Plan. In 1940, only 9.8 percent of Arkansas farms had electrical service. Eventually, Arkansas Power agreed to work with the federal government to bring power to the rest of the state's farmers. For a $1 down payment, the government would deliver electricity and wire a small house. The program was a success. By 1950, 66 percent of the state's farmers had power. By 1960, it was 97.5 percent.

The expansion of electrical service overseen by Couch was expensive. To encourage Wall Street to lend the utility money, the state protected Arkansas Power by making it a monopoly. In return for building the grid, Arkansas Power was granted exclusive operating rights. The new state regulators carved up all of Arkansas into power franchises, from the dominant Arkansas Power to smaller electric co-ops. To protect customers in these competition-free zones, the Arkansas Public Service Commission, which took over the duties of the Department of

Public Utilities in 1945, regulated power prices. All across the United States, power companies and local lawmakers struck similar deals.

Once Arkansas Power received a monopoly franchise, the rules made sure no one else could follow. To obtain a certificate of public convenience and necessity to be a utility in Arkansas, a corporation needed to be in the business of "owning or operating" power equipment. But it couldn't own or operate power lines unless it was a utility. There was another catch. To be a utility in Arkansas, you had to serve customers. But if you were trying to break into the state, you didn't have any customers. It was a catch-22 of Harvey Couch's devising. At his urging, the state had successfully dug a moat around his utility, and hadn't built a drawbridge.

This configuration remained in place when Skelly eyed Arkansas as a route. The state had no idea what to do with Clean Line because it had never seen anything like it before. The officials who wrote the 1935 law didn't anticipate anything remotely like what Skelly was proposing. "The reason it is silent is because merchant transmission providers are a new creature. I mean, obviously, they weren't contemplated when the '35 Act was passed," Public Service Commission staff attorney Valerie Boyce testified. Without any rules, the state regulatory body struggled. "The commission has never faced this issue before," Public Service Commission chairman Paul Suskie said at the time.

Clean Line was stuck. It wanted to build and operate a power line. To do that, it needed permission to build facilities and to be recognized as a utility. But under state law, it did not qualify as a utility because it didn't own a power line (or power plant). At a hearing in December 2010, Colette Honorable, a member of the Public Service Commission who would go on to serve on the Federal Energy Regulatory Commission, zeroed in on this problem. It was a "chicken and egg paradox," she said. Which would come first? Should Clean Line go through the time and expense of getting environmental permits and easements to build some electrical equipment, so that it could claim it owned assets and apply to be a utility? Or should it get the stamp of approval from the state to be a utility first? It was unclear how it could sign up the customers it needed to become a utility without being a utility—and what did that even mean if it intended to zip through Arkansas on the way to Tennessee?

"How do they get to that point if we won't allow them to move forward?" Honorable asked Emon Mahony, assistant attorney general. "Well," he responded, "it is kind of a little bit of a difficult situation." They should push ahead and get permission to build the power line, and then come back later for permission to be a utility, he suggested.

"It's a high risk, wouldn't you agree?" she asked.

"It is a little bit of a risk. I'll agree with you," the young lawyer, only a couple years out of law school, responded.

Both Skelly and Jimmy Glotfelty also testified before the commission. Glotfelty urged commissioners to approve the project. This would allow both wind developers in Oklahoma and the TVA in Tennessee to start taking them seriously. Skelly warned that if the commissioners delayed, the project would sink into a "cloud of ambiguity." How were Clean Line employees supposed to go out and talk to the public, lawmakers, and environmental groups without being able to tell them that the state was willing to at least consider the project? What would happen if they spent months talking to landowners, figuring out the route, and then were told the state wouldn't allow it? "We would have wasted not only a lot of our time and money, but we would have wasted a lot of time from the folks that we had asked to participate and give us input on the proper siting of a line like this," Skelly said. Arkansas was asking Clean Line to spend years and perhaps millions of dollars to design the transmission line without giving a sense of whether the idea would be stillborn. Harvey Couch had been dead for nearly seventy years, but the bureaucratic trap he set years earlier was still set.

A month after the hearing, the Public Service Commission issued its ruling. The commissioners noted that Clean Line wasn't asking for any money from Arkansas or its electric customers. And it said that "Clean Line's efforts are laudable and its work is to be commended." But the law was the law. And the lawmakers hadn't contemplated that a private company would come into Arkansas to build a power line to move inexpensive wind power. Clean Line did not meet the definition of a utility and it could not proceed with its plans.

It was the first time in his career that Skelly had tried to develop a project that crossed state lines. He had learned an important lesson. There really is no national grid—or federal grid oversight. The contiguous United States might as well be forty-eight different grids, each overseen by an individual state and state officials, looking to protect their local interests.

If Arkansas was frustrating for Clean Line, Oklahoma dashed any remaining hopes that the company could sashay through the regulatory process. Two weeks after filing an application in the state, Oklahoma Gas & Electric filed papers to oppose the project. Clean Line was not planning to sell electricity in Oklahoma, it argued, therefore the state had no authority to approve it as a utility. (But unless it was a utility, it couldn't build a transmission line.) It was another catch-22.

Pretty soon, an association of small, independent oil companies entered the fight against Clean Line. Behind closed doors, a crowd of lawyers hammered out a deal. Clean Line had hoped to begin work soon, but would have to wait for sixteen months until late 2011 to get final approval. Under the deal, Clean Line got its utility status, but as a concession it agreed to a more arduous process to use eminent domain. If Oklahoma Gas & Electric or any other state utility built a power line, they could use eminent domain as long as they successfully argued that the line was in the public interest and would serve Oklahomans. Skelly could argue until he was blue in the face that adding inexpensive wind power benefited everyone by adding low-cost energy that didn't emit carbon dioxide, created jobs, and developed

Oklahoma's maximum potential for energy development, but it didn't matter. It was not in the narrowly defined public interest.

This rejection meant the success of Clean Line rested on a green light from the federal government. But the federal authority, spelled out in Section 1222 of the Energy Policy Act, was untested. It explicitly said the government could partner with a private transmission builder, sharing its power of eminent domain, if the project served a purpose beyond enriching its owners. Even as the matter wended its way through administrative law hearings in Oklahoma, Skelly turned his attention to Washington, D.C. Striking such a first-of-its-kind deal would mean scrutiny and lengthy legal agreements. But the negotiations were worth it, even essential. Skelly didn't want to use eminent domain, but he also knew that without it, it would be difficult to erect a one-hundred-mile-long line of towers strung with direct current lines—much less one across two entire states. Without eminent domain, the power lines might end up looking like lightning bolts that zigzag across the land, inefficient and possibly prohibitively expensive.

All Skelly needed the Department of Energy to do was follow its own rules. Except there were no rules. That a private, merchant company would try to build a transmission line was unprecedented. This realization dawned on Skelly in waves. First, he couldn't figure out who to talk to at the Department of Energy. No one seemed to know which office would handle it. "If you go to FERC [the Federal Energy Regulatory Commission] and say I want to put a pipeline in, they go 'Good. Talk to the guy in Room 359.' The DOE didn't have that," he said.

Glotfelty, as a former high-ranking federal energy official, used his contacts to arrange a meeting with Holmes Hummel, a senior policy advisor in the department. She was working on the Obama administration's push to bring in more renewable energy. Her reading of annual reports from the national laboratories convinced her that wind costs would continue to fall. She wanted to find a way to get more wind onto the grid. But politics were getting in the way. "We were facing a significant block of resistance to a clean-energy transition as a national policy from delegates in the American Southeast," she said. A federal mandate to get a certain percentage of electricity from clean sources, an idea under consideration, would be hard on the region. If it tried to rely on its meager local wind resources, higher power prices would be the result. What's more, it could require shutting down local power plants, which were important sources of jobs and political patronage.

Politicians from the southeastern states, such as Mitch McConnell from Kentucky and Alabama's Jeff Sessions, were aggrieved. "They felt there was an undue burden on the region for lack of natural endowments. We saw that there was a large amount of low-cost resources just a reach away," Hummel said. Even before Skelly showed up, she was looking for a way to get wind from the Great Plains into the Southeast.

Skelly and Glotfelty thought the meeting went well. There was interest in their project inside the department. They thanked Hummel for her interest. At the time, policy required that noncitizens had to be escorted around the building. But Skelly and Glotfelty, both citizens, could wander the halls of the Department of Energy. Skelly tried to buttonhole people in the hallways and poked his head into offices. "Hi, we're working on this thing that we're really excited about," he said. That policy has since been changed.

Skelly's and Glotfelty's wanderings paid off. In the summer of 2010, the Energy Department issued a request for proposals, soliciting interest from private companies to build new transmission lines. Skelly wanted to make an impression at the Energy Department and make sure they took the company's proposal seriously. He ordered up boxes made from recycled aluminum with the Clean Line logo— two overlapping lines like hanging power lines—etched onto them. "I want it to be impressive and I want it to be memorable and I want it to look like we're serious, like we have the money to do this," he told staff. The gimmick worked. Years later, energy officials would joke that the Energy Department building could collapse and workers picking through the rubble would come across the Clean Line proposals inside an intact aluminum box.

Clean Line spelled out in the proposal what it wanted to do. "The most vexing challenge blocking continued growth in the renewable energy industry is the expansion of the U.S. electric transmission grid. The existing transmission system was primarily built as a result of local utility planning—connecting population centers with nearby fossil fuel power plants. In the last decade, the nation has expressed the desire to move away from fossil fuels and towards a clean energy economy," it read. To get there required a new approach.

The line itself would cost between $3.5 billion and $4 billion, but Clean Line made clear it wasn't asking for any money from the government. It would be financed privately. Building it would open up a vast new energy resource in the United States, like building the Trans-Alaska Pipeline had opened up the North Slope's oil reserves. The energy targeted was Oklahoma wind. The state itself could produce in excess of 267,000 megawatts of relatively steady wind, ten times the amount consumed by the power grid covering Oklahoma, Kansas, and Nebraska, as well as parts of Missouri, Texas, and Arkansas. Build one line and wind developers will flock to western Oklahoma. "Plains & Eastern will make possible some $12 billion of renewable energy projects that otherwise cannot be built due to limitations of the existing grid," the company said in its proposal.

The aluminum boxes arrived on Tuesday, July 6. The capital had celebrated the nation's birthday with its usual fireworks on Sunday and then taken Monday off as a federal holiday. If there was any excitement that the United States was moving ahead with an innovative power project, a giant 4,000-megawatt high-voltage direct current line, it was snuffed out by the end of the week.

On Friday, the Xinhua Chinese news agency issued a news item. The State Grid Corporation of China announced that a new high-voltage direct current line was operational. It could carry 6,400 megawatts of power from the Xiangjiaba Dam and hydroelectric plant on the Jinsha River along the southern edge of Sichuan province. Carried on a set of two lines, the power flowed 1,230 miles eastward to Shanghai, the largest city in the world and the country's financial hub. Shu Yinbiao, the deputy general manager of the State Grid, said that it could supply fully one-third of the city's power demand. Compared to that Chinese project, the Plains & Eastern was a piker. It was significantly shorter and would carry less power. The Xiangjiaba line was lauded as the longest and largest power line ever built. It crossed through eight Chinese provinces and had taken three years to permit and build.

There was zero sense of urgency in Washington. The Clean Line proposal sat in the Energy Department. As Clean Line filed briefs in Oklahoma and Arkansas, nothing seemed to be happening in Washington. According to an undated Energy Department memo, proposal review was "put on hold" while federal officials discussed whether the government should follow the law and use its power of eminent domain. Letters were sent to various companies—such as TVA and Entergy—to gather input.

In addition, the Clean Line proposal was basically homeless. No one inside the agency knew who would pick it up. "Nobody was focused on it," said Lauren Azar, a Wisconsin state regulator who became a special advisor in the department in the summer of 2011. Her mandate was to identify and remove barriers to develop electrical infrastructure. Skelly visited her a few weeks after she began work. She looked into Clean Line's application before the meeting and realized it had entered a bureaucratic deep freeze. "There was not really much happening with their application," she said. She hoped to change that.

Skelly was pleased and sent her a copy of the languishing proposal. "The private sector has the resources and the desire to invest in our aging infrastructure," he wrote, "and we respectfully ask that the DOE exercise its authority to make it possible." He was on bended knee. Clean Line had waited more than a year for the department to begin to pay attention to it. In the meantime, Clean Line had submitted another application—this time for its Grain Belt project in the Midwest.

In April 2012, Clean Line got some good news. Daniel B. Poneman, a deputy secretary and one of the top officials at the department, sent a letter to Clean Line that laid out some conditions before moving ahead. Clean Line needed to reimburse the government millions of dollars for its work on the project and assume all liabilities. Moreover, Clean Line had to agree to use eminent domain "only as a last resort."

If Clean Line agreed, the federal government was ready to move ahead and begin a lengthy environmental review of the project and community outreach. This wasn't a final approval, but it was good news. Oklahoma was also on board, but Arkansas still refused to given Clean Line the authority to build a power line.

In February 1959, Soviet engineer Andrei Mikhailovich Nekrasov gave a well-attended talk about direct current power lines at a winter meeting of the American Institute of Electrical Engineers. Speaking before a couple hundred people in the ballroom of the old Statler Hilton hotel in midtown Manhattan, Nekrasov described several existing lines, including one in Moscow and another in Sweden that fed power to an island.

What captured the attention of the attendees—and most likely of the Central Intelligence Agency spies in the audience—was his description of an experimental direct current line that would carry power from the giant Volga Hydroelectric Station, then under construction, to the industrialized region of eastern Ukraine. (The CIA had a thorough file on Nekrasov, although his activity seems limited to publishing numerous articles on the Soviet power system, making him by all outward appearances an engineer interested in exchanging ideas.)

"It may be considered that the problem of DC power transmission will be essentially solved by constructing the Stalingrad–Donbas line," Nekrasov said. This line would move about 750 megawatts from the dam to a converter station about three hundred miles to the east. It would dwarf the existing Moscow direct current line and other small-scale direct current lines. Just as the Soviet Union had leapt ahead in the Space Race, Nekrasov promised it was ahead with new ways to move electricity.

Americans were not interested in coming in second again. They had already looked skyward at an orbiting Soviet satellite. They did not want to follow the Soviet lead in the development of a new power grid to move electricity over hundreds of miles. The U.S. government wanted to build a direct current power line in the United States that was bigger and longer than the Soviet's. The question of where to put it didn't take long to determine.

The winter before the Russian paper was presented was particularly mild in the Pacific Northwest. This meant less power demand for heating. There was also an economic recession that dampened the call on aluminum from the region's smelters. This spelled problems for the Bonneville Power Administration. A federal agency created to market power from dams on the Columbia River, the BPA had been generating financial surpluses since its inception decades earlier. Now it was operating at a deficit.

The BPA had excess power in the summer, when snowmelt swelled the river. But the Pacific Northwest had mild summers. Its power demand peaked in the winter. Southern California was the opposite. During the summers, large numbers of air conditioners, a relatively new and rapidly spreading consumer item, drove up power demand. In the winter, power demand was much lower. Numerous coal- and gas-burning power plants sat idle during the winter months.

What if a trade could be arranged? In the summer, BPA could ship some of its excess power south. In the winter, Southern California could ship some power north. A month after John F. Kennedy's inauguration, the new president directed his interior secretary to figure out if this trade made enough sense to build a power line.

By the end of 1961, a task force had examined the financials and declared that the Pacific Northwest and Southern California should be connected "at the earliest practicable time." A DC line connecting the Columbia River and Southern California would make use of excess power and help the BPA's finances. Moreover, this new connection, soon to be called the Pacific Intertie, would be bigger and longer than the Soviet's DC line from the Volga dam to the Donbas region.

There was opposition from governors, senators, and private utility companies who didn't want competition. The Pacific Northwest worried that Southern California would take all of its power, and insisted on legislation that ensured only surplus power could flow southward. Advocates such as the California Department of Water Resources, which needed the power for its irrigation pumps, talked about a national solution. "State boundaries should not be permitted to form a bar to sensible and economic utilization of electric energy resources," R. C. Price, the acting director of the department, wrote in a letter to the head of the BPA.

The Pacific Intertie would be "an effective challenge to the Communist bid for supremacy in the development of high-voltage long distance direct current transmission of electricity," wrote a nonprofit corporation owned by rural electric cooperatives that favored the project. A newspaper journalist, after talking to the Kennedy administration, made a similar point. "The proposed California line would outmatch the Soviet Union, which has been experimenting with such systems as a part of its efforts to expand its power production rapidly and overtake the United States by 1970," the article stated. The Pacific Intertie was a cog in the machine that would deliver the U.S. a victory in the Cold War.

An appeal to patriotism didn't prevent some pitched fights. California's large private utilities complained the government shouldn't spend tax dollars on the project. Political leaders in north-central California worried that the line would benefit BPA to the north in Oregon and Washington States and the City of Los Angeles in the south, but not them in between. Eventually, traditional alternating current transmission lines were expanded to carry power from the Columbia River dams south into California's Central Valley. In the Pacific Northwest, congressmen insisted on a guarantee that California couldn't use the line to deprive it of the inexpensive electricity used to attract investment and economic development. If this protection was included, Washington governor Albert D. Rosellini promised to drop his opposition.

In September 1963, President Kennedy traveled to Hanford, Washington, to help inaugurate a nuclear power plant. Speaking at the event, he asked for a show of hands of people born in the state. Then he asked people who weren't born in Washington to raise their hands. It is not recorded how many people raised their hands, but enough of them did to enable Kennedy to make his point. "When we develop these resources in the Northwest United States, it is just as well that the country realizes that we are not talking about one State or two States or three States; we are talking about the United States," he said.

He also urged support for the Pacific Intertie. "We must construct an efficient interconnection between electric systems . . . and between regions," he said. Three months later, Congress approved $8.5 million in funding to begin work on the line. By then, Kennedy was dead.

Even after Congress approved spending to begin buying equipment, fighting continued. Congressional Democrats charged the new administration of Lyndon Johnson with moving too quickly and giving too much away to private developers. Local officials were eager for low-cost power from the Northwest. The head of the public service department in Burbank, California, lobbied Congress directly. It bought natural gas from Texas to burn and generate its own electricity. The mere threat of a power line had led gas suppliers to cut their rates.

After several more months of wrangling, the sides struck a deal in July 1964. In an act of political appeasement, Congress passed a bill that required the Pacific Northwest get first call on the power. The Interior Department issued a gushing press release. The direct current power line was part of a new interconnection that would carry enough power to be "equal to the daily power needs of six San Francisco's or five Washington D.C.'s." Construction began almost immediately. The Pacific Intertie began near the Washington-Oregon border, crossed into Nevada, and headed southeast over the desert. Years in the future, revelers on their way to Burning Man would cross under the double wires about twenty-five miles before entering Black Rock Desert. It was the last tall structure they would see before encountering the spindly figure that gives the bacchanalia its name. The lines headed east of Reno, then south for a couple of hundred miles before taking a dogleg to the west and ending at a converter station in Los Angeles, where the direct current would be changed back into alternating current and distributed.

When it opened, in 1970, the *Los Angeles Times* said that it "ranks with the pyramids as an engineering feat." At 865 miles, it was more than twice the length of the Soviet line. At 1,440 megawatts, it could convey nearly twice as much power as the Soviet line (and it would be upgraded over the years to carry 3,800 megawatts). Floyd E. Dominy, then the head of the federal Bureau of Reclamation, boasted the line would be "America's first and the world's largest. This project will make our nation the world leader in an exciting new transmission technique."

The line was dedicated in September 1970 and operated without problems for six months until a 6.6 magnitude earthquake struck underneath the converter station in Sylmar, California. After repairs, the line went back into operation in 1973. The Pacific Intertie cost about $700 million. The federal government paid $300 million. The rest came from the Los Angeles Department of Water and Power and several private utilities, including Pacific Gas & Electric and Southern California Edison. A Swedish company, later part of the industrial conglomerate ABB, developed the technology. It considered some of the equipment so cutting-edge and secretive that Swedish engineers assembled it in a huge locked room.

President Johnson, speaking at a "Victory" breakfast for one thousand people in Portland not long after the project was approved, called attention to how the project had overcome inertia and politics. "You have proved that if we turn away from division, if we just ignore dissention and distrust, there is no limit to our achievements," the Texan declared.

Forty years later, in 2015, Skelly drove under the Pacific Intertie. Seeing the line inspired Skelly. It was proof that years of political fights could be overcome and projects built. He thought about the years of debate: Should it be built? Was it necessary? Should it be privately owned? Publicly? All of that was forgotten. Ultimately, they decided to build it and today it is one of the backbones of the grid in the western United States. All those air-conditioned Hollywood studios? They could thank the Columbia River, nearly a thousand miles to the north.

By the time Skelly saw it, the Pacific Intertie had been operating for nearly half a century and likely had at least that much more life in it. "If politicians and engineers could come together to build it, there was hope for the Plains & Eastern, the Grain Belt, the Rock Island, and the Western Spirit," he thought.

12

Julie, Virgil, and Janey

D ave Berry, Clean Line's head of strategy, held a marker in his hand. Spread out in front of him was a large map of the United States' electrical grid. A group of employees huddled around the map. Bob Marley gazed down on them from a black-and-white photograph on the wall.

Clean Line employees gathered in the conference room regularly. Weekly staff meetings were opportunities for anyone to speak his or her mind in what were often freewheeling discussions about politics, corporate strategy, and even the company's funding. This time, they were planning the route for another giant transmission line. The plan was to tap into the wind in the southwestern corner of Kansas, where prairie breezes were stiff and steady. Dodge City is one of the windiest cities in the country; it is no more than 110 miles northeast of Guymon, Oklahoma, where the Plains & Eastern would begin.

Berry led a conversation that weighed the pros and cons of different options. The Grain Belt line needed to end at a point on the grid that could serve both as an on-ramp for the delivery of 3,500 megawatts as well as a hub for easy distribution of power to neighboring states. Walmart and Amazon locate their distribution warehouses off major interstates, not two-lane country roads. Berry was looking for something equivalent on the grid. After a lengthy back-and-forth, Berry and the employees settled on a location south of St. Louis, more than five hundred miles to the east of Dodge City. Berry uncapped the marker and drew a line from Kansas up to Missouri.

Diana Rivera, the former Harvard MBA and one of the company's first employees, watched. As the marker squeaked across the map, she thought how she would soon be on the road, calling on county commissioners and other local leaders, along the line Berry traced. A few months later, Clean Line learned the substation near St. Louis couldn't handle that much electricity. So, on paper, the line extended further east. It now stretched nearly eight hundred miles to Terre Haute, Indiana.

The company had started with an idea in Skelly's mind of transmission lines shooting out from the middle of the country, like spokes on a wagon wheel. One of the biggest challenges was to turn his ideas into physical realities. These lines needed to cross interstates and farmers' fields. They needed locations for thousands of towers. These were linear projects. If a line went into a valley, it had to come out of that valley and connect to the next valley. Plotting workable routes would take months and involve countless community meetings. And every tower meant a new set of neighbors would have to be notified and mollified.

In November 2012, Clean Line kicked off a series of community meetings for the Plains & Eastern transmission line in Van Buren, Arkansas. At 7 a.m., Clean Line opened the doors at a county extension service office to anyone who wanted to come and learn about the project and offer an opinion. Clean Line held well over a hundred such meetings in local chamber of commerce conference rooms and community college lecture halls. The company provided urns of coffee and light refreshments. Long maps unfurled across tables. Skelly, Hurtado, and other Clean Line employees introduced themselves to attendees and discussed the maps and the project. Soon, groups would cluster around satellite photographs discussing river crossings and cemeteries.

Julie Morton saw the announcement for the Van Buren meeting in her local newspaper. She woke up feeling sick that morning and didn't attend. But she wanted to know more. She had held many jobs, including running a small candy store at a mall and reporting the weather on a local CBS affiliate. For the last few years, she had worked as a right-of-way agent. When the Ozark Gas Transmission Pipeline was developed, she talked to landowners in order to secure easements to put the pipe under their land. She switched later to working for Oklahoma Gas & Electric to help route transmission lines. When she first heard about the Plains & Eastern, she was in her early sixties. She was still full of piss and vinegar. "I'm a tough-hided old girl," she said, describing herself. She had been married twice—once for three years and once for three months.

She prided herself in understanding her corner of the world. "Arkies," she said, "we're as poor as dirt, but we have land, and that is our core. When you start messing with that, you are messing with our heart and soul." For her, owning land was about holding on to something precious. "Jeff Bezos and them have all the money, but they don't have all the land," she told me.

After missing the first meeting, Morton called Glenanna O'Mara, the mayor of Cedarville, a hamlet just north of Van Buren where Morton lived. Did she know about the transmission line? Had she attended the meeting? The mayor said she hadn't attended either. The mayor demanded another meeting, this time in the one-story brick-sided Cedarville City Hall. Mario Hurtado and about ten other Clean Line representatives came out and set up what was becoming a routine presentation: topographical and satellite maps unfurled on tables and posters tacked up that showed their monopole design.

About a hundred people showed up. Some talked favorably about the positive economic benefits. Mayor O'Mara applauded Clean Line.

"The company was good to come here," she said. For Skelly, these meetings were a project imperative, not a kindness. "You have a duty to talk," he said. "You have a duty to tell the public what you are doing because you are asking public officials to make decisions with respect to the public interest." Clean Line created a spiral-bound book, exactly one hundred pages long, with maps and drawings of what the poles would look like, along with answers to questions. (A sampling: "Who owns Clean Line Energy?" "Is wind power cost effective?" "What happens when the wind stops blowing? Do the lights go off?") Clean Line distributed the books to interested public officials. In addition to the printed material, Skelly and other Clean Line officials hit the road. Where other corporate chief executives might take to the lecture circuit, proclaiming themselves as green gurus at industry conferences, Skelly made the rounds of cities and states on the route of the lines. It wasn't glamorous but it was necessary.

Before setting off for this round of meetings, Skelly scored an improbable victory. Houston is the international capital of hydrocarbons. In this city whose economy is built on the demands of the internal combustion engine, Skelly led a bond campaign to borrow $166 million to build new bike paths and green spaces. During the day, he worked on transmission lines. After work and on weekends, he focused on greening Houston.

Skelly had been a bicyclist since he was a teenager. He grew antsy if more than a few days passed without a spin on his bike. Over the years, he had explored Houston's bayous on two wheels, discovering paths that often stopped abruptly and required a detour onto streets. The bond money would connect these paths and turn Houston into a more outdoor-friendly city. To the chamber of commerce crowd, he argued that young professionals wanted this kind of recreation. Without it, he told a panel on Houston's future, "they just won't move."

Throughout the fall of 2012, he had pitched a vision of a network of bike paths throughout the city. When the money was spent, more than 75 percent of the Houstonians would be able to get into downtown without getting in a car. "For a city that is thought of as a concrete jungle, this was a big deal," said Luis Elizondo-Thomson, who Skelly recruited to be the campaign director.

The bond was a long shot. The November ballot was crowded, and Skelly and others working on the campaign worried that debt fatigue or confusion would hurt his chances. Voters were asked to borrow money for firehouses, community colleges, and schools. But when the votes were counted, more than 68 percent of voters approved the bike measure. It got more votes than any other bond issue.

If the bond distracted Skelly, it didn't show. The same month as the election and the first community roundtables in Arkansas, Clean Line got a financial boost. National Grid, one of the largest British companies listed on the London Stock Exchange, said it would invest $40 million in Clean Line. Not only would they take an equity stake in the company, they purchased the right to buy a large stake in the transmission projects. National Grid executives praised Clean Line's "portfolio of compelling projects that will advance the growth of renewable energy and the modernization of America's energy infrastructure." The deal was a much needed capitalist vote of confidence in Clean Line. No longer merely a scrappy group from Houston, they now had the backing of one of the world's largest electricity companies. Moreover, National Grid valued the company at a significantly higher level than when the Ziffs and Zilkhas had initially seeded it.

Skelly was overjoyed. The money and support boosted his vision of a new electricity infrastructure. Plains & Eastern was moving along, and routing for Grain Belt was under way. Two more lines were in development. Not all would work out, he figured, but if just one or two could be built, Clean Line would be a success. Each of the four was enormous. Any of them could carry "the energy equivalent of three Hoover Dams," he said while announcing the National Grid investment.

When Julie Morton attended the Cedarville meeting and learned of Clean Line's plans, she was less exuberant. Years of working as a land agent had made her proficient at reading maps. At the Cedarville meeting, she traced the two main proposed routes. Both ran like arrows across the flatlands of Oklahoma and then began to wiggle north and south, as the foothills of the Ozark Mountains made straight lines difficult. Her fingers followed the lines eastward, across the border into Arkansas, and she saw the northern of the two routes—Alternative 4-D—would run right through Shadow Mountain Village.

Her father had developed the rural subdivision in the early 1990s, plotting out large, reasonably priced lots. He had kept the two best lots, the ones that backed up to the hollow that ran down to Cedar Creek, where bats swooped out of the dogwoods and ate so many mosquitoes that you rarely had to put on any repellent. He had passed away, but his widow still lived in a house on one of these lots. Julie Morton lived next door in a two-story house built on the other lot.

She drove home from the meeting angry. Her anger focused on Mario Hurtado, the highest-ranking Clean Line official at the meeting. She parked in front of her garage, got out of the car, and walked past the raised beds that were full of heirloom tomatoes and peppers every spring, and let herself into her mother's house. She plopped down at the kitchen table and began to cry.

"Mother, this is going to ruin us," she said. After sobbing for a few minutes, her head cleared. Years later, she described the moment. "I said, 'Mother these things are hard to stop.' And then we set to work."

———————

Julie Morton turned out to be an effective adversary and a tireless political agitator and organizer. She started driving around western Arkansas in her red Nissan pickup truck, talking to state legislators and county commissioners. Her aim was simple: kill the Plains & Eastern. She often wore small diamond studs in her ears and held her head tilted slightly backward so her chin poked out. She brushed her black-tinged-with-white hair out. And when she spoke, she often stretched out her words in a way that left no doubt she was a local. She understood the economic troubles that rural residents faced and shared the stew of resentments they felt. She could tap into the distrust of outsiders coming into Arkansas, antipathy toward the wealthy, and anger that she and her neighbors were being asked to make a sacrifice so someone else could turn a profit. She helped spread the word eastward, forging alliances with people such as Alison Millsaps, who regarded Skelly not as a renewable energy pioneer but a businessman with all-too-sharp elbows. "To us, it looks like Big Oil is moving over and here comes Big Renewable," she said. Like Julie Morton, she also became consumed with the fight. Her ten-year-old son gave her a Valentine's Day card one year in which he wrote: "Mom worries about me and my sister and Clean Line."

Millsaps's distaste for the fat cats she spied hiding behind Skelly motivated her. Why should her neighbors be asked to "give up their land for the greater good?" she asked. "We shouldn't be asking them to give more money to the Ziffs and the Zilkhas." This message became a rallying cry. At one meeting, protesters waved about a hundred chartreuse signs with black lettering: "No eminent domain for personal gain."

Clean Line dispatched a string of bright young employees into Arkansas—and later Missouri, Indiana, and elsewhere—to talk to locals and political leaders about the benefits of the project. By and large, they did a good job of winning over allies. But they were outsiders, both geographically and culturally. Most had moved around, from city to city, in search of the best education, jobs, and professional challenges. Their attachment to any particular place was weak. They had this in common with Clean Line's leadership. Skelly had moved around: Virginia to Costa Rica, then Minneapolis, back to Costa Rica, and then Houston. And he was the child of immigrants: his father had left England and Ireland to find a better opportunity at a factory in Erie, Pennsylvania. Mario Hurtado's family came from Colombia.

Michael Zilkha moved from London to New York and then Houston. Jayshree Desai, the chief operating officer at Clean Line, was the daughter of Indian immigrants.

Julie Morton was much more rooted. She had been born in Hope, Arkansas. "Like Bill Clinton," she said. Her family went back a few generations in the state. She told me that the best way she could describe Arkies' love of the land was to quote the scene from *Gone with the Wind*, when Scarlett O'Hara's father talks to her about the importance of Tara. I later looked up the quote: "Land is the only thing in the world worth workin' for, worth fightin' for, worth dyin' for, because it's the only thing that lasts."

Morton talked to every justice of the peace she could reach in eastern Arkansas. She argued against what she saw as Clean Line's unnecessary land grab. The justices sit on a quorum court in each county, determining taxes and expenditure plans along with the county judge. Some looked favorably on the project. It would bring jobs and create opportunities during construction. There would also be lease payments to landowners to obtain easements, and Clean Line promised to pay each county $7,500 for every mile of line. "While these payments are not required, it is one way of demonstrating our commitment to being a good citizen," Clean Line wrote to local officials. Under federal law, if the Department of Energy agreed to work with Clean Line, the federal government would end up owning some of the line, making it exempt from state and local taxes. Clean Line said it would make payments to compensate for the lost taxes.

This financial arrangement made sense to many people in Arkansas. Greg Bell, owner of a pharmacy in Hector, where customers could buy fishing supplies and examine guns for sale while filling their prescriptions, worried the local school was at risk of closing. Hector needed economic development, and if that was what the Plains & Eastern would bring, then he was for it. Sure, Clean Line needed some land, but how was that different than building the interstate or nearby Lake Dardanelle. "Any time there is progress, there has to be concessions for progress," he told me.

The Sierra Club has been a longtime opponent of long-distance transmission lines because, in the 1970s and 1980s, they were built to create new markets for coal. But after a lengthy internal discussion, and with significant internal opposition, the group backed the project. They opposed coal, natural gas, and nuclear, Sierra Club executive director Michael Brune explained at a conference. "We have to be in support of something," he said. The Arkansas State Chamber of Commerce backed the line, calling it an innovative project that would bring low-cost energy to the state. And, it pointed out, big manufacturing and tech companies wanted to build where they could get the kind of inexpensive, clean energy that the Plains & Eastern would deliver.

Morton was tireless in her opposition, even when it was no longer personal. Clean Line scrapped the Alternative 4-D route that ran through the subdivision her father built in favor of another several miles to the south. She didn't slow down. She gathered six thousand signatures in 2013 and 2014. Whatever benefits would flow from construction of the line "would not come anywhere close to being satisfactory compared to the permanent devastation left in the wake of a transmission line of this magnitude," the petition stated. Those signatures were just the beginning, she said. She gathered names and built a political organization—and then began applying pressure on Arkansas's congressional delegation. "If there is one thing those son-of-a-bitches understand, it's votes," she said. Congressional aides started coming to rural meetings and told her to have her allies stop calling. They got it.

Clean Line responded by holding more meetings. It operated under the premise that Plains & Eastern would help the region by providing low-cost, inexpensive energy. All it had to do was get out into communities and tell its story. Clean Line wanted to convince

Arkansas of the merits of the project, not win at any cost. "Just because you're a decent businessman doesn't give you the right to not be a decent person," Hurtado said.

When the Department of Energy began an environmental review of the project, it brought in Jane Summerson, a government environmental compliance officer, to run meetings to gather public input. Clean Line footed the bill. She said she was "astounded" that Clean Line was willing to have so many meetings. Federal law requires one meeting, she said, during the environmental review. Clean Line had fourteen. Clean Line was allowed to consult with Oklahoma tribal leaders by teleconferences. Company officials held numerous face-to-face meetings. Years later, she described working with them as rewarding. "They were ethical and honest and willing to go way beyond what was required," she said. Summerson had run several previous environmental reviews for the federal government. Nothing in those projects compared to the anger she sometimes faced at meetings in Arkansas. And her job was to gather facts. She couldn't speak. It was like being awake and alert during surgery, but unable to move or react.

After the meetings were all concluded, Summerson got a personal email one day from Alison Millsaps. "I know we were rough on you," she wrote. "Thank you." Summerson began to cry.

Julie Morton was one of the roughest on Summerson, the entire process, and Clean Line. At a public meeting in Van Buren in 2015, she railed against a for-profit corporation taking land. "This is a slippery slope down which our government is taking us. Where will it end? Will the corporations end up owning all the property and there will be no private property rights left?" she asked. "A word to the wise, if you keep trampling on the rights of we the people, you may have another American Revolution on your hands."

Jayshree Desai encountered this anger up close. It was company policy that everyone needed to go out and attend meetings, and that included Desai. She would talk to locals about how the project would bring renewable energy and help everyone. "And they look at you and see you have a nice house in Houston, Texas, and a nice frickin' car," she said. She realized that when many people looked at her, and her co-workers, they saw a bunch of elites from Houston trying to impose their business plan, their climate solution, on rural Arkansas.

The majority of people at Clean Line felt their mission was important, and they were uncomfortable being seen as greedy intruders. No one had a good way to resolve this disconnect. It left Desai feeling conflicted. "We are asking people for a sacrifice," she said. "And what are we sacrificing in return? I don't have a great answer for that."

———————————

On a January day with clear blue skies, I met Morton in Van Buren. I left my car in a Walmart parking lot and climbed into her red Nissan pickup. It had more than 100,000 miles on it. The windows were so dirty that I couldn't tell if they needed a wash or if smoke from a nearby forest fire was making it hard to see. The fabric on the ceiling of the cab was missing, revealing patches of ocher-colored padding.

She drove me to a treeless, half-complete subdivision of brand-new $1 million homes. It was on a hill and the transmission line was to cut through an adjacent wooded valley. The doctors and lawyers who bought these homes would be able to look out and see the 110-to-150-foot-tall towers at eye level. "It was utter panic," she said. After a spin through the subdivision, we headed a couple miles north. Past a lumberyard, she pointed out the local offices of the Arkansas Valley Electric Cooperative. "I want to show you our little solar farm and how innocuous it is," she said. We pulled into the parking lot of the co-op and on a hill behind the building were two large clusters of solar panels. It was located next to an existing substation—and didn't require any new overhead wires. "They hooked it right in and no devastating transmission lines were needed," she said.

Morton said she thought the co-op's solar installation was how energy should be done. Local power for local people. Later, I looked up the plant's details. It sat on one and a half acres and could generate 500 kilowatts. That could handle, maybe, 150 central air-conditioning units humming away, maybe 200 if they are particularly efficient. The Plains & Eastern would deliver 14,000 times more power. Put another way, you would need to build twenty co-op-sized solar arrays every mile from the Oklahoma panhandle into Memphis. There was no comparison. When I asked Morton about this, she shrugged it off. It's a start, she said.

We headed out, passing a lot of small farms and country homes. We worked our way east. Morton talked at a rapid clip about western Arkansas and the people she had met. At one point on our tour, on Interstate 40 west of the town of Ozark, we came around a curve and encountered a beautiful vista. For five, or maybe ten miles, were rolling wooded hills. Morton grew excited and took her hand off the steering wheel and swept it above the dashboard to make sure I saw the carpet of trees. "Right through this beautiful valley is where it would have come," she said, asking me to imagine what the Plains & Eastern would look like once built.

A few seconds later, her Nissan sped under a power line that crossed the highway and dove down the hillside into a valley on our right. It was easy to see, sitting in the middle of a hundred-foot right-of-way that had been cleared of all plants taller than a foot or two. A series of large wooden "H"-shaped towers marched into the distance.

I asked her if she opposed the line we had just passed. The poles didn't bother her. "Those aren't as intrusive as a metal tower," she said, "and, of course, it is serving people." What she meant was that it was part of the alternating current network that provided juice to local homes. Plains & Eastern was shooting power to Memphis and beyond. It wasn't serving her people, at least not anyone nearby.

As we kept driving, I started thinking about what was so different between the Plains & Eastern DC line and the AC line we had just driven under. I wondered if Morton's fight wasn't about the line, but about power. Not the electrical kind, but political and financial power.

Who has it? Who wants more? And who is left feeling that they are just places on an empty map to the financial bigwigs of faraway cities?

Sometimes these feelings about being powerless are enough to stir up anger, and sometimes they can stir up a genuine revolt. I asked Morton if she had ever heard about the time in the 1970s when rural opposition to a power line metastasized into an armed uprising. She laughed and said she didn't know anything about it. "So, I wasn't the first?"

The battle took place on the black-soil plains of western Minnesota, when farmers took up arms against a high-voltage direct current line being built from the coalfields of North Dakota to the Twin Cities. The farmers used what they had at their disposal: snowmobiles, tractors, CB radios, and, eventually, guns.

It is largely forgotten today, even in history books of radical movements of the era. A local political scientist professor named Paul Wellstone called it "America's Energy War." This was years before he mounted a quixotic, but ultimately successful, campaign to unseat an incumbent U.S. senator and became one of the most progressive voices in Washington, before dying in a plane crash in 2002.

Virgil Fuchs helped lead the revolt. Now retired, he no longer lives on his farm. He splits his time between a home on the shore of the glittering Lake Koronis, a few hours west of Minneapolis, and a winter home in Phoenix. He has done well in his life, and, curiously, he owes much of his success to the power line he fought against.

Well into his eighth decade, Fuchs is ramrod straight, tall, and lean. His bushy white sideburns spread out from under the temples of his glasses. Nearly forty years after his fight, his distrust of outsiders hasn't receded. The first time I met him, after parking in his driveway and introducing myself, he stared at me and asked: "Do you work for the power company?" His wariness is understandable. During the battle over whether the power line would be built, a private investigator said the power company had hired him and off-duty policemen to infiltrate the opposition and sow paranoia.

Virgil and Janey Fuchs were raising five children and farming the muddy soil of Stearns County, Minnesota, when they first heard about plans to build a high-voltage direct current line. The line crossed their farm.

In the early 1970s, U.S. energy production stopped growing and the era of oil imports began. Government officials in Washington feared energy shortages. So did two power companies that supplied power to the Fuchs farm and its neighbors. In 1972, the companies joined together to build a large power plant that would eliminate existing and projected power shortfalls. But what type of power plant? Oil wouldn't do. That would leave the farmers at the mercy of the Middle Eastern emirs. Nuclear took too long to build. By the time a new nuke was operational, the rural power companies would have gone through several years of deficits, buying power on the regional grid at potentially exorbitant prices. The only good option was coal. And a few hundred miles to the west, in Underwood, North Dakota, was 225 million tons of lignite coal buried under a few feet of soil. It was, according to promotional material the companies put together, "the largest supply of low-sulfur lignite in the world."

Once they settled on coal, the Minnesota power companies had a choice. Should they move the coal by train to a new power plant near the Twin Cities? Or should they build the power plant at the North Dakota mine—and move the electrons by overhead wire? New state laws would make it difficult and time-consuming to build a new coal plant in Minnesota. They needed the power soon. So, by process of elimination, they arrived at a plan. They would build a power plant next to the mine and ship the electrons over a 430-mile DC line.

They applied for and received $537 million in loans and guarantees from the federal government's Rural Electrification Administration, the federal agency that had helped bring power to rural farming communities since the Franklin Delano Roosevelt administration. As the project was developed, costs more than doubled. The loan officers at the Rural Electrification Administration kept issuing more loans until the power line became the largest project it ever financed.

The plan called for about 1,650 towers to hold the wire aloft from North Dakota into central Minnesota. The towers resembled 150-foot-tall Erector Sets in the shape of men, with four legs and two arms stretched out with teal-colored glass insulators holding the lines. These steel men would stand out in the farmers' fields, and many of the farmers felt their land and lifestyle was secondary to the growing power needs of the state. They weren't entirely wrong.

Virgil Fuchs was in a tractor, working in one of his fields, when the surveyors parked their trucks on the road alongside his farm. They set up orange cones to warn the practically nonexistent traffic on the un-improved township road. The men were there to calculate longitude and latitude. They wanted to put a transmission tower on his land.

This was early June 1976, a month before the United States cel-ebrated its bicentennial. Fuchs was thirty-four years old, a German American Catholic whose family had fought in the Civil War to earn a homestead in western Minnesota. He couldn't understand why the line had to run through his property. The power companies had first talked about routing it a few dozen miles to the south. This route would have taken it through the Bonanza Valley, where the soil was inferior. "Sand farmers," he called his neighbors. His homestead was "nice, flat, highly productive agricultural land," he said.

To determine the route, a consultant for the United Power As-sociation, one of the companies building the line, imposed a grid on the land between North Dakota and suburban Minneapolis. It assigned the boxes a number from zero to six, and programmed a computer to calculate a route that avoided higher-numbered boxes. Highways were assigned a four, forests a three. Lakes were elim-inated from consideration. The consultants assigned farmland a zero. With these inputs, the computer spat out a route that was almost entirely farmland.

Some towns gamed the system. For obvious reasons, federal aviation rules prohibited power lines within three miles of airports of a certain size. Tiny Starbuck, Minnesota, announced a major expansion of its airport to exceed the federal size requirement. Never mind that no one in Starbuck was clamoring to turn the turf airstrip into a mini-O'Hare. The town said it intended to build a much larger lighted runway. The power line was routed north of Starbuck. Later, plans for the new runway were shelved.

On that summer day in 1976, Fuchs watched the surveyors set up tripods and tribrachs. He grew angry and something in him snapped. He turned the wheel of his tractor and drove straight toward the surveyors. He crashed through the two-stranded wire fence, down into the drainage ditch that separated the field from the road, and up the other side. He would have collided with the surveyors' pickup truck if one of them hadn't jumped in and driven it away. Then Virgil Fuchs drove his large farm tractor over the surveying equipment. He turned the tractor around and drove over it again. He said it felt cathartic. He then drove up behind the pickup truck that a supervisor was in and rammed it. Fuchs yelled that if they ever came back to that area, he would wreck their equipment again and "shoot all of them," one of the surveyors later recounted. Fuchs drove away, leaving the stunned surveyors in his wake.

A sheriff's deputy drove to the Fuchs farm that evening and told Janey that Virgil needed to report to the county courthouse in the morning. Word about what happened passed by phone among the farmers. When Fuchs reported to the courthouse, dozens of enraged farmers gathered outside. Like Virgil, they felt like their land was seen as little more than a route to deliver electricity to someone else. The power companies were frustrated. They had followed the rules. For two years, they had submitted applications, obtained certificates of need from the states, and held meetings. And just as construction began, farmers were running over their equipment. Inside the courthouse, Virgil Fuchs was arrested and charged with two felony counts, arraigned, and released on bail that afternoon.

Similar to the federal siting process that Clean Line would go through three decades later, the Minnesota law required power companies to identify a broad corridor and then a route within that corridor. They notified every landowner in the corridor. After months, the farmers started attending meetings and asking questions. The process, designed to create public acceptance, had ended up angering a large number of farmers. Virgil Fuchs spent a lot of time before and after his arrest driving around organizing his neighbors. The result was a bucketful of acronyms: Keep Towers Out (KTO), Counties United for a Rural Environment (CURE), Towers Out of Pope [County] Association (TOOPA), and the General Assembly to Stop the Powerline (GASP).

Fuchs said he disliked how the power companies "told us to take it and like it." More than anything, he didn't like feeling like his livelihood or his family was being taken for granted. Art Isackson, a farmer interviewed in 1978, pointed out that the Minnesota Department of Natural Resources didn't want the line run over wooded areas because it could break up animal habitats. "I guess a skunk is worth more than a farmer," Isackson said.

A week after Fuchs's arrest, surveyors continued mapping the route. This time, about fifty farmers showed up to get in the way of the surveying equipment. It was the beginning of the Energy War, or at least a rural rebellion. For the next few months, disruptions continued to escalate as surveying continued. Sheriffs sometimes arrested farmers who blocked the surveyors. Sometimes the sheriffs refused. As tensions rose, the new governor, Rudy Perpich, ditched his bodyguards, who warned him against going to the farming communities, and drove to Lowry to meet with the restive farmers. He tried unsuccessfully to convince them to participate in arbitration.

By the end of 1977, protesters were obstructing survey crews across Pope and Stearns Counties. State troopers and local sheriff's deputies scuffled a few times with the protesters. Then, in January 1978, the situation deteriorated even further. The sheriff's departments in these rural counties were outnumbered. One sheriff wrote to the governor that his department has "only two deputies and can by no means control a crowd as we had today."

The protesters pushed for a moratorium to stop construction. Governor Perpich had other plans. He appeared on a small television station that served western Minnesota and said he had "gone the extra mile" to defuse tensions and resolve the matter. The time for discussion had ended. The power line was lawful and would be built. He then announced the largest mobilization of state troopers in Minnesota history. Perpich dispatched 225 troopers, half of the state's garrison, to Pope and Stearns Counties. This occupying army arrived with enough matériel to quell a riot: twenty-nine quarts of tear gas, thirty-one quarts of smoke, gas masks, bullhorns, riot batons, and hundreds of flex cuffs, as well as the state helicopter.

The show of force allowed the surveying work to finish, but the farmers' anger found other means of expression. They hung a banner across the street in Lowry that referred to the governor. In spray-painted letters, it read: "Rudy's Rangers Go Home No Powerline." A stuffed figure in maroon, the color of the troopers' uniform, hung in effigy next to the sign. In March, someone shot out the windshield of a parked truck where a security guard was keeping an eye on equipment. He was cut by glass shards.

With the spring thaw, work on the towers began. On May 17, a line of cars and trucks carrying fifty protesters descended on one construction site. "We didn't want no bloodshed and we told them we didn't want no trouble and we just wanted to leave," Michael Lee Smith, a worker, told police afterward. Smith and the six other workers were allowed to walk away. Then the group picked up large rocks to try and crack the concrete foundations for the newly erected tower. They attacked a thirty-five-ton crane nearby, smashing its headlights and windows, slashing tires, and punching holes in the radiator. As the group left, one woman grabbed a five-gallon container of motor oil and emptied it in the cab. Then the caravan drove off.

The first tower fell in early August 1978. Someone removed the bolts at the tower's legs. Before the end of the month, three more towers toppled. Later that year, another tower fell. Two legs had been cut with hacksaws, putting pressure on the remaining legs, which bent under the stress. There was a break during the winter and then, in March 1979, another tower fell. Police noted the snowmobile tracks leading up to the tower, but were unable to follow them anywhere that led to an arrest.

In the late evening of December 30, 1980, a bomb threat was called in to the Lakeside Ballroom in Glenwood. Sheriff's deputies went to investigate and found no sign of a bomb. While law enforcement was in town, someone took a blowtorch to a tower about eight miles to the northwest. The two legs on the southern side were cut about three inches above the concrete foundations. The two northern legs couldn't hold up the 150-foot tower and it fell. Throughout the spring, another fourteen towers fell. Glass insulators that dangled from the towers' arms and held up the wires were often used for target practice.

No one was ever arrested for the destruction of the towers. Virgil Fuchs says it wasn't him and he doesn't know who was responsible. "I can tell you honestly, I don't know who did it," he said. Then he smiled. "I have a suspicion, but he's dead." At the time, a mimeographed newsletter suggested slyly it was the work of "bolt weevils."

Toppled towers were rebuilt and eventually the project was completed. The Energy War ended in defeat for the project's opponents, who went back to farming. One day in July 1979, without any announcements, the North Dakota–Minnesota line was energized. Years later, at a symposium held at the Pope County Historical Museum, Trooper Phil McMahon recalled the confrontations. He was based in Pope County and was asked to arrest his neighbor—"people I knew, people I respected." Wearing an ironed and starched shirt, he appeared nervous at the symposium. He had a steno pad full of notes about what he wanted to say, but didn't glance at it.

"Our grandparents found it hard to accept the railroad going through, which sometimes took their lands and went right through where their buildings were," he said. After the railroads, he said, came the freeways. Routing them also required splitting some farms and taking some farmland. "Whenever there is progress, there is change and change does not benefit everyone. Change is hard for some people to accept."

———————

Virgil Fuchs drove me out to a bean field he once farmed. We parked the car and got out to walk under the line. The power company towers are the tallest structures, man-made or natural, for miles around. The line stretches out into the distance. It is windy, but every so often the wind dies down. During the quiet stretches, Virgil Fuchs says you can hear the power overhead. It sounds like bacon sizzling on a skillet.

I asked if it still upsets him to see the line. "It's here. What are you going to do," he responded. He thinks some good came of it. The community got closer and stronger. Janey Fuchs wasn't so sure. She said the line bothers her every time she drives past it. "There are no good memories there," she said. She keeps glass insulators left over from the construction of the power line on her dining room table. Glass doesn't conduct electricity, so they were used to connect the line itself to the steel towers. She had baked several insulators in her oven at 300 degrees and then dumped them in ice water. The rapid temperature change caused them to crack. The thick glass was a crazy quilt of cracks. She placed an electric candle underneath. It was her act of revenge. She had taken part of the power line and turned it into art. She kept a couple and gave a set to each of their five children for Easter.

After the power line was connected, Virgil decided he didn't want to spend much time on the portion of his land under it. So he borrowed from the bank and bought more land and larger farm equipment. That way, he figured, he could spend less time on any one part of his land. This turned out to be a smart investment. He rented most of the land to other farmers and has retired handsomely. "The power company made us rich people," he chuckled.

The Minnesota DC took seven years to turn an idea into reality. Big energy infrastructure projects take time. Two years earlier, the giant Alaska crude oil pipeline opened its valves to swallow the first barrels of warm crude. It was nine years after earth-moving equipment began clearing a path for it. A federal injunction had halted the pipeline at the legal request of environmentalists. Congress had to get involved, in the anxious days after the 1973 OPEC oil embargo, to push through the project.

The Plains & Eastern and Clean Line's other lines were funded privately. Its investors did not have infinite patience. Once under way, the project needed to establish and maintain momentum. This attracted suppliers and encouraged public support. Julie Morton understood that one way to effectively attack the line was to slow down its approval and deprive it of precious momentum. Skelly and his team pushed the project through various regulatory agencies and met with local leaders. Julie Morton and her confederates looked for ways to whip up dissent and use that to bring pressure on politicians.

Skelly was trying to prove there was a way to combat climate change that could also generate a good economic return. Prove that, and capital would be unlocked to fund more projects. It was a virtuous cycle of capital and climate change. Morton intended to interrupt and slow down this cycle until it collapsed.

13

The Week from Hell

B efore any of its giant transmission lines could be built, Clean Line needed to assemble certain ingredients: there had to be a company that wanted to generate and ship electricity, a buyer on the other end for the electricity, a slew of permits, and enough money to make payroll until all these pieces could be assembled.

Finding companies to build wind farms was not going to be a problem. Wind developers were like a class of elementary students waving their arms in the air to get the attention of the teacher. In August 2013, Clean Line issued a formal request for information to wind developers to gauge interest in the Plains & Eastern. The response was heartening. Fifteen wind developers put their hands up to use the line. Added up, they wanted to develop 17 gigawatts of wind near the substation that would serve as the on-ramp, four times the planned capacity.

A trial balloon for the Grain Belt yielded similar results. The Kansas-to-Indiana line would carry 3,500 megawatts. When Clean Line tallied the responses, developers expressed interest in building 13,500 megawatts.

In May 2014, Plains & Eastern held an open solicitation—the next step after a request for information. Clean Line said it expected wind farms to charge about $20 per megawatt for power. Add another $20 for transmission and the Plains & Eastern could deliver power into the South and Southeast for $40 or so for every megawatt hour, enough to power 750 homes for an hour. It was, Michael Skelly said, "a compelling value proposition."

Renewable energy prices had fallen sharply. In the first months of its existence, Clean Line had sent an unsolicited proposal to the TVA. It said it could deliver megawatts "at an all-in cost of under $70 per MWh." Five years has passed and the delivered price of wind power had been cut nearly in half. The turbines themselves were getting bigger and more efficient at catching the wind's kinetic energy and turning it into electricity. The industry was getting better at figuring out where to place the turbines. The result was that turbines could run more hours of the day, generating more power for less cost.

If things looked good on the seller's side of the Plains & Eastern, there was still work to do on the buyer's side in Tennessee. Atop Skelly's to-do list was to convince the Tennessee Valley Authority to become the anchor customer for the power. And that meant developing a relationship with the relatively new head of the authority.

Bill Johnson was a large presence: six feet, five inches and more than 275 pounds. A former Penn State offensive lineman, he was an imposing presence. He came to the TVA in January 2013 when his climb up the corporate ladder ended in a stunning public humiliation. After getting a law degree, he worked as a lawyer for several years before joining Progress Energy, a North Carolina electricity company. A fan of the Grateful Dead and reggae, he was once advised he needed to get rid of his beard to help his career. From that point onward, he kept a clean chin. He had made it to the top of the company when cross-state rival Duke Energy proposed a merger in 2011. Duke's CEO Jim Rogers proposed that Johnson would run it.

The merger closed at 4:02 p.m. on July 2, 2012. As planned, the fifty-eight-year-old Johnson became the CEO. At 5 p.m., the board met in executive session. When they came out, they asked Johnson to resign. His tenure as the leader of the largest utility in the country lasted a little more than an hour. Jim Rogers would be taking over.

A week later, the silver-haired Rogers testified before the North Carolina Utilities Commission. It wanted to know what the hell was going on. Sitting upright in a dark blue suit and red tie, Rogers explained that as the merger talks progressed, board members lost confidence in Johnson. They didn't think he had done a good job overseeing Progress's nuclear plant. Members of Duke's board also worried Johnson kept his board of directors in the dark. Duke executives feared they would suffer retaliation if they dissented from their new CEO's decisions. They felt Johnson's leadership style was "autocratic and discouraging of different points of view," Rogers said. Bill Johnson, through his lawyers and later in testimony, said this was all an attempt to smear him. Jim Rogers had engineered a boardroom coup, he countered. His complaints came too late. Johnson was out of a job.

A few months later, a search firm contacted him about taking over the Tennessee Valley Authority. He negotiated with the board for a salary that would pay him nearly $1 million a year with significant bonuses for reaching certain incentives. He began work in January 2013 in TVA's headquarters in Knoxville. By the time the federal fiscal year ended on September 30, 2013, he had taken home $5.9 million, including bonuses and deferred compensation. It made him the highest-paid federal employee in the country. Not bad for nine months' work.

Created in the 1930s, the TVA was one of many innovations by President Franklin Roosevelt to lift the country out of the Great Depression. The government corporation would provide low-cost power to the impoverished Tennessee Valley. Eighty years later, the TVA no longer received government disbursements. The rates it charged for power supported its operations. But the federal government still appointed its board of directors and approved its debt levels. Providing inexpensive power to nine million people in Tennessee and neighboring states remained at the heart of TVA's mission. And if it ever strayed, local government, community, and business leaders weren't shy about reprimanding its executives at one of its regular public meetings.

In May 2014, TVA held a board meeting in Memphis. Randy Spicer, the controller at a nearby Nucor Steel plant, spoke. The company had restarted the plant on the Mississippi River in 2006, turning out steel shaped into crankshafts and oil pipelines. But Nucor's steel competed in a tough global marketplace. Keeping the plant open required producing steel at competitive prices. And that meant it needed inexpensive electricity.

"I want you to recognize how important energy is to our operation," Spicer told the TVA's board of directors. "Electricity is one of the highest costs in our operation. It is our lifeline. We are not making steel without it." The plant employed 425 people, who made an average of $75,000 a year. If TVA didn't keep power costs low, the majority of those workers would head to the unemployment line. Spicer's point was that TVA was inextricably wrapped up in the economic health of the community and the welfare of its people.

Johnson had his marching orders. Keep power prices low. Keep the lights on. Protect the environment. Avoid the kind of massive coal ash spills that had befouled tributaries of the Tennessee River in 2008. When he took over as CEO, a quick survey of his new TVA empire presented some stark facts. Nuclear plants provided the cheapest electricity at $6 per megawatt hour. But building a new nuke was a colossal headache. One of Johnson's priorities was to finish a "new" nuclear plant called Watts Bar Unit 2 near Knoxville. Construction on it had begun in 1972, before it was placed in a long-term hold. When it was finished in 2016, after a forty-three-year gestation, costs had risen fifteen-fold to $6.1 billion. The next cheapest source of TVA's electricity was coal at $32 per megawatt hour and then natural gas at $39.

The message Skelly delivered to Johnson as often as he could was that electricity from the Plains & Eastern would be cost competitive. It was no longer some extravagance that would cause Nucor Steel or Memphis community activists to wave their fists. The price of wind power was falling. TVA's own long-range planning called for adding more renewable energy as long as it was cost effective. But whether TVA would agree to buy power from the Plains & Eastern was only partly a decision driven by dollars.

———————————

A few days after the Memphis meeting, a letter landed on Bill Johnson's desk. It was from Senator Lamar Alexander and Representative Stephen Fincher, both powerful Republicans in Tennessee. The letter asked several pointed questions about the Plains & Eastern power line. Would it be more expensive than other sources of power? Isn't it unreliable? Is TVA's demand for power growing? Wouldn't it require eminent domain? The letter was addressed to Johnson and his board of directors. The message was clear. Powerful politicians were taking a close interest in the Plains & Eastern.

Johnson's reply was both courageous and diplomatic. The first point he made was that federal law guided his decisions. "TVA is mandated by Federal statute to provide electricity at rates as low as feasible," he wrote. His second point was that while TVA had been in discussions with Skelly, Clean Line had not yet presented a proposal. There was nothing to act on.

The rest of the letter gently pushed back against some of the lawmakers' misconceptions. Wind couldn't be dispatched like a coal plant, but utilities across the country were integrating more wind onto the grid successfully. "The power grid is a complex, interconnected network of generating plants, transmission lines, and distribution facilities," Johnson wrote. It was and would remain a balancing act, but the TVA was up to the challenge.

Then he turned to answer the question about whether TVA needed new generation. TVA was expected to experience slow growth, Johnson wrote, which didn't require adding a giant new wind resource. But there were other changes afoot. "TVA has retired or will retire a substantial portion of its coal fleet," he wrote. So, as TVA closed older plants, there would be need for new sources of power. He also downplayed eminent domain. Clean Line was negotiating deals with landowners and wouldn't need to knock down any homes. Moreover, the land could still be farmed or ranched, as long these activities don't interfere with the line operations. Johnson was pushing back against the politicians. But he was also conceding that if TVA were to buy wind power from Plains & Eastern, it would likely mean retiring some older coal plants.

Lamar Alexander's intense dislike for wind power wasn't new. The first evidence of it was in 1999, not long after he ended an unsuccessful run for president. After spending the better part of a year in Iowa, he returned to Knoxville to visit family, childhood friends, and acquaintances from his days as president of the University of Tennessee.

At the time the TVA was building the first wind farm in the southeastern U.S. on a mountain north of the city. It could barely be called a wind farm. There were only three turbines, but they were visible from some parts of Knoxville. Alexander decided he didn't like them. They were "100-yard-tall, monstrous structures," he said later, and didn't belong on any scenic Tennessee mountain.

Alexander and his wife, Leslee, bought undeveloped property on Nantucket Island, off the Massachusetts coast, in 2001. It was a beautiful place to build a vacation house. Across a small street, a road led down to the beach. Nothing obstructed the view of the Atlantic Ocean. The week they closed on the property, news broke about Cape Wind, a plan to build 170 wind turbines in the middle of Nantucket Sound. Alexander wasn't happy about this wind farm either.

In 2002, Alexander reentered politics. He ran for and won a U.S.

Senate seat from Tennessee. He inserted himself into the long-running, contentious, and emotional debate over Cape Wind. One of the bills he proposed in his first year aimed at requiring local governments to sign off on any offshore wind development. The bill went nowhere.

He asked the TVA to place a two-year moratorium on new wind farms, even though the agency didn't have plans to build any. Stephen Smith, director of the Southern Alliance for Clean Energy, a watchdog group, was dumbfounded by Alexander's campaign. He described it as "the most direct assault on wind power we've ever seen by a United States senator."

Alexander picked up the effort again a couple of years later. He introduced a new bill that would give a local veto over any wind farm—and required any offshore wind farm to go through an extensive environmental assessment that duplicated already existing efforts. On the Senate floor, he reminisced about his grandparents' farm, which had a small windmill that pumped water from underground. "The windmills we are talking about today are not our grandmother's windmills," he warned.

He implored his fellow senators to imagine a wind turbine placed in the middle of Neyland Stadium at the University of Tennessee. "It would rise to more than twice the height of the highest skybox. Its rotor blades would stretch almost from 10-yard line to 10-yard line. And on a clear night, its flashing red lights could be seen for 20 miles." When a reporter asked him if his opposition to wind was based on his undeveloped property in Massachusetts, he didn't deny it. "I wouldn't like seeing them wherever my house might be and I don't want other people to have to look at them either," he said.

A few years later, President Obama appointed a critic of the TVA named Neil McBride to serve on its board of directors. He went to Washington for his confirmation hearings. Ushered into Alexander's office, the senator showed him a Photoshopped picture of the wind turbine poking out of Neyland Stadium. It appeared as if it had been built on the fifty-yard line.

This is to scale, Alexander told him. The message was clear: Alexander had never encountered a wind farm he liked. McBride would have to contend with this dislike from 2011 until 2014, while he was a member of the TVA board. "The fundamental issue that TVA had to deal with, when it started looking at wind power, was Senator Lamar Alexander's deep opposition to wind power of any kind and any place," said McBride. McBride was keenly aware of the importance of not angering him. "You just want to keep him happy," he said.

Skelly met with TVA head Johnson several times in 2014 and felt like he was making progress. In November, Johnson wrote an official letter to indicate TVA's interest in the project. The federal government was pushing hard to clean up existing power plants, he wrote, and having access to the country's best wind could come in handy. The contents of the letter fit squarely with Skelly's pitch. The Plains & Eastern would deliver the wind potential of the Great Plains right to Memphis's doorstep. Johnson's letter confirmed he had faith in Clean Line.

More good news followed as the year ended and 2015 began. Illinois approved the route of the proposed Rock Island line. The federal government issued a draft environmental impact statement for the Plains & Eastern that was overwhelmingly positive. It wasn't a final decision, but it was a critical step toward a decision. Tennessee also gave approval to Clean Line to build the Plains & Eastern in the state. With Oklahoma also on board, all that remained was Arkansas. State leaders were being intransigent and Clean Line was banking on the federal government agreeing to participate in the project. That would make it a project of national importance and Arkansas wouldn't be able to block it.

There were still brushfires. The quorum court in Pope County issued a proclamation against the transmission line. The line would "be an enduring eyesore to Arkansas . . . with little positive effect," the thirteen-member court determined. Resolutions like this popped up across Arkansas. Julie Morton had passed around proposed language, then gathered up the resolutions, and presented them to the Energy Department. At her urging, Pope County judge Ed Gibson called Arkansas senator John Boozman to convey the decision of the local government.

On the same day as this proclamation, in early 2015, Clean Line submitted its first official proposal to TVA. The proposal came from Invenergy, a Chicago renewable energy developer. Invenergy had cut a deal with Carroll Beaman, the octogenarian who had driven around to visit his neighbors and gather up wind leases on thousands of acres of land in the Oklahoma panhandle. The bowlegged former oilman had amassed a land position and then brought in a company to figure out how to turn it into a giant wind farm. Beaman had the family connections to get people to sign leases. Invenergy had the capital. It had built fifty-six wind farms, worth about $15 billion, as well as gas-fired power plants across North America and Europe. But it had never attempted anything on the scale of what Skelly envisioned.

The proposal to the TVA was rudimentary. Invenergy offered to sell either 500 megawatts or 1,000 megawatts at a set price. The line could carry about 4,000 megawatts, but the extra wouldn't be a problem. TVA had a fabulous transmission grid. It had once planned to build more than a dozen nuclear power plants, and had built up the copper wires to move all that power around the Tennessee Valley. But it had scrapped the nuke plans not long after the partial meltdown at Three Mile Island.

The TVA's transmission system was "gold-plated," Skelly said. TVA wasn't using the transmission system to anywhere near its full capacity. "And it belongs to the federal government, by the way," he said. This was federal property sitting around unused. Skelly figured he could sell the TVA as much renewable energy as it wanted, and TVA could use its transmission system to move the rest to other parts of the grid.

Dave Berry, Clean Line's head of strategy, spelled this out in testimony before the Tennessee Regulatory Authority. Once the power was delivered into Memphis, it could be moved in several directions. It could head into Virginia, North Carolina, or into the Southeast. It could be sold to Duke Energy, the company that Bill Johnson once thought he would run; or into Georgia, Mississippi, and Alabama, territory controlled by the Southern Company; or up into the Northeast, where a large power market and grid covered a swath of the Mid-Atlantic states.

The proposal was well-received by TVA officials. Bill Johnson had asked his staff to contrast the Clean Line proposal with some local solar developers who were also dangling some low-cost power before the TVA. Berry and Jayshree Desai flew to Chattanooga to meet with TVA officials and reported back what they felt was good news. "Bill Johnson is clearly pushing the organization to engage with Clean Line," Berry reported back to Skelly.

With everything clicking, Skelly was working on making sure Clean Line had enough money. He had been talking with Global Infrastructure Partners to provide a third round of funding—each round valuing the company higher than the last. GIP would be a perfect partner. It invested in infrastructure such as airports, wind farms, pipelines, and ports around the world. It was accustomed to spending large sums on projects that paid off for decades.

For Clean Line, February 2015 looked to be a month full of hope and promise. The company was closing in on additional funding. The federal approval process seemed to be moving forward. And in the middle of the month, two Clean Line employees were getting married. They had met at the company and gotten engaged. But the premarital bliss was interrupted in what came to be known around the Clean Line offices as the "the week from hell."

On Thursday, Arkansas's two senators, John Boozman and Tom

Cotton, introduced legislation aimed at Plains & Eastern. They called it the Assuring Private Property Rights over Vast Access to Land Act, or the Approval Act for short. The law would require the federal government to get approval from the governor and public service commission of each affected state before it could use eminent domain to build a transmission line with a private company. Thus, Arkansas would have a veto over the Plains & Eastern.

"When a road, pipeline or power line is built the use of eminent domain is sadly unavoidable in some cases," Boozman said. "However, this difficult decision should not be in the hands of Washington bureaucrats. If a project is not good for Arkansas, our governor or public service commission should have the power to say 'no.'" Skelly was furious and flabbergasted. "We were blindsided," Skelly said. "Holy shit, here we have these guys who are pro-business, running on pro-jobs platforms, and they are just sticking it to us."

Jimmy Glotfelty was in Charlotte, North Carolina, when he heard about the bill. He grabbed the first flight he could to Washington. The cofounder and head of Clean Line's governmental affairs was stunned. He had thrown a fundraiser for both Cotton and Boozman and their staffs hadn't even given him any advance warning. On his way to the capital, he got into what he called a "verbal spat" with one of Cotton's staffers. The staffer told him the legislation wasn't aimed at Plains & Eastern. Glotfelty was apoplectic. The joint press release mentioned only one project by name: the Plains & Eastern.

In Houston, Sarah Bray's phone was ringing nonstop. Reporters wanted a reaction from Skelly and were besieging the Clean Line spokeswoman. She looked around and the members of Clean Line's management she could see were "freaked out." But she couldn't find Skelly anywhere. On a hunch, she went upstairs to Michael Zilkha's offices on the nineteenth floor. Zilkha had a terrace that wrapped around two sides of the building. Skelly was there stewing.

Skelly and Bray crafted a response that emphasized the positives without antagonizing the senators. "We believe in the merits of the project, to bring manufacturing jobs to Arkansas and low-cost energy to the state. We're following all of thc stcps required," Skelly told reporters, "and ultimately, the decision-makers will decide." There was no sugarcoating it. If passed, the bill would kill Skelly's plans and turn Clean Line into a husk. If that wasn't bad enough, it felt downright tribal to Skelly. The United States power grid was hopelessly balkanized. Grid operators and utilities had their fiefdoms and had little to no incentive to cooperate with each other. There were all kinds of legacy institutional barriers to building transmission lines that crossed from one fiefdom to another. This bill would further etch that balkanization into law. Skelly was trying to build a transmission line that connected different grids and tied them together, sharing power in a way that created stability and enabled clean, inexpensive power to flow across state lines.

This bill seemed likely to keep everyone looking out for their own parochial interests. It allowed every state to put up a barrier until it got what it wanted. Skelly was trying to build an interstate and Senators Boozman and Cotton wanted to allow every state to put up tollbooths.

The next day, the Iowa Utilities Board handed down a decision that was a setback for the Rock Island transmission line. Things weren't better in Tennessee. The TVA issued a planning document that was lukewarm toward wind and long-distance transmission. This last item seemed to come out of nowhere. Just a few months earlier, a top TVA official had told Glotfelty that TVA's purchase of power from Clean Line was a matter of when, not if. In meeting after meeting, Clean Line officials had left TVA offices in Chattanooga barely able to contain their excitement. TVA told them they were sizing up a new gas plant in Memphis so that it could better accommodate wind from Oklahoma. Companies wanted to bring data centers to the Tennessee Valley and TVA officials said they were excited about using wind power to close the deal. There hadn't been any hint of hesitation—and now TVA had issued an official resource plan that talked about wind in the 2020s, but not sooner. It made Skelly wonder what they had been talking about for the past eighteen months, and whether he and other Clean Line people had gotten too enthusiastic.

The Saturday wedding—between two project developers—was to be a nice break from a long week. The couple, Ally Smith and Daniel Hodges-Copple, wed at a Methodist church in Houston. As the pastor told the couple, "This will be a day you remember for the rest of your lives," a college friend of the groom jumped up and yelled that the church was on fire. Indeed, at the top of the ceiling, electrical wiring leading to suspended lights had caused a beam to catch fire. Sparks were falling on the second and third pews. Everyone evacuated and the wedding continued on the lawn. Photographs of the wedding party feature Houston fire trucks.

A few days later, the Arkansas senators' proposed Approval Act scared off a Clean Line investor. A couple of managers at Global Infrastructure Partners called Skelly. "They said to put the pencils down. We are done. This isn't going to work out," he said. "Investors hate regulatory risk. And we served up a plateful that week."

// **14** //

This Is Not Your Ordinary Transmission Line

I n late April 2015, Skila Harris, a senior advisor to United States Energy Secretary Ernest Moniz, briefed her boss on the Plains & Eastern line. He would soon need to make a decision, and his staff wanted to make sure he knew what was at stake. A nuclear physicist, Moniz was known for his memorable head of hair that looked like George Washington's on a good day and the psychotic assassin's from *No Country for Old Men* on a bad day. He was very good at avoiding Washington D.C.'s pitfalls. He would need these navigation skills for the Plains & Eastern.

One of his options was to partner with Michael Skelly and Clean Line, get the line built, promote renewable energy, and encourage private markets to build a new generation of transmission wires. This choice had risks. If Clean Line botched the line and it turned into a boondoggle, the Obama administration would be dragged into it, even though it was privately financed. There was a bigger risk, involving eminent domain. If Arkansas farmers protested, or worse, began an armed confrontation to stop construction, the government would be implicated.

Moniz's other choice was to turn the application down. But this path also carried consequences. It would spell the end of Clean Line and send a chilling message to pension funds, university endowments, and others considering investing in large renewable energy projects. And if there was going to be an energy transition, it was going to involve a huge amount of capital. The message Moniz would send to investors would be that they should look elsewhere. Approving the line would help create a new architecture for moving power, he thought, and rejecting it would put them "back to square zero."

Moniz's top aides had heard from lawmakers on the project. The loudest voice had been Congressman Steve Womack's and he was not happy. He was a Republican who represented Fort Smith and other parts of western Arkansas. Details of the meeting were relayed to Moniz at the briefing, which was held around a large table in a conference room adjacent to his office with a view of the Smithsonian Castle and oil paintings on loan from the nearby National Gallery of Art. "The project had received significant congressional interest," Moniz was told, according to notes from the meeting.

Skila Harris, a Kentucky native and former director of the Tennessee Valley Authority, brought Moniz up to speed on the project. It was working its way through a lengthy environmental review, she said, which should be wrapped up in the fall. The questions they needed to address before making a decision were whether the project was in the public interest, whether it was technically and financially viable, and what benefits it would bring to each state it traverses.

That last item was problematic, she said. Clean Line had proposed building the wind farm in Oklahoma and sending the power through a 720-mile direct current line to Tennessee. In between, Arkansas received little. Harris said this was politically untenable. "It poisoned the well," she said. Congressman Womack, in a letter earlier that year, had fumed that the line treated Arkansas as a "highway for power in

Tennessee" and offered "no guarantee that Clean Line will supply power to my constituents and my state."

As Moniz began to grapple with the issues, he realized a political solution was needed. The first step was to make sure Arkansas got power from the line. This requirement was conveyed to Clean Line, and Skelly had relented and agreed to add a converter station in Pope County, Arkansas, that would serve as a 500-megawatt off-ramp. It added to the project's overall cost, but was necessary to make the project acceptable.

Initially, Clean Line had decided not to put it there for a good reason: Skelly and others worried it would upset the powerful Arkansas utilities. "If you made the utilities mad, they would make your life hell," Jimmy Glotfelty explained. It was a damned-if-you-do, damned-if-you-don't situation. Provide power to Arkansas and upset the established utility companies. Don't provide power, and upset the politicians.

Harris also briefed Moniz on the issue of eminent domain. Before Moniz had become secretary, the government had sent a clear message to Clean Line. It could pursue its application, but the federal government was not planning to give the company carte blanche to seize land for their line. In a letter to Skelly, Energy Deputy Secretary Daniel Poneman had been clear. "Eminent domain authority would be used only as a last resort after negotiations in good faith have concluded," he said. That was just a letter. It would all have to be spelled out in a legally binding contract. All of the promises that Clean Line had made, and then some, needed to be hammered out.

The Arkansas delegation continued to fume about the project. They called Moniz to Capitol Hill for a meeting where the state's two senators and four congressmen had expressed their firm opposition to the project. The energy secretary played the role of wily arbiter. He committed to the elected officials that he would make sure Clean Line agreed to the various compromises, including limits to eminent domain and making sure contracts and jobs to build the line went to

Arkansas companies. Until they did, there would be no federal participation. But once Clean Line agreed, he would approve it. "I was prepared to take as long as it took to have the right project," Moniz said.

Over the next several months, in meetings in Washington and
over the phone, representatives of the Energy Department and Clean
Line worked on a deal. They traded copies of the proposed terms and
conditions. Harris and Moniz wanted assurances that Clean Line
wouldn't make a halfhearted effort to negotiate with landowners and
then blaze a line of condemnation claims across Arkansas. There
needed to be conditions in the deal so that Clean Line used eminent
domain as sparingly as possible.

Harris spent a year, along with department lawyer Sam Walsh,
pushing Clean Line. The final document reflects the effort. The department didn't want to be left as the owner of a half-built project in
case there was an unexpected bankruptcy, so Harris and Walsh negotiated an escrow account with money to deal with abandonment.
They pushed language requiring Clean Line to meet landowners in
person, and spelling out efforts to track down long-lost relatives who
might own a one-eighth stake. No detail was too small for negotiation.
The Energy Department sought and received final approval over a
standard script to be used by right-of-way agents when they met with
Arkansas residents.

The promises Skelly had made about paying counties in lieu of
taxes (the land would be owned by the government, and therefore
exempt from taxes) were put into the legal document. For hosting
thirteen miles of line, Van Buren County got a $99,000 one-time pay-
ment and $5.7 million over forty years. For 27.6 miles, Pope County got
$207,000 and $36.8 million. Altogether, Clean Line agreed to nearly
$150 million in payments. And that didn't include what would be
needed to lease the actual land.

The biggest issue was eminent domain. For Harris, it was a personal issue. Her grandparents' house in Bowling Green, Kentucky,
had been bulldozed to clear room for Interstate 65. One of the road's

cloverleafs took drivers right through where the house once stood. "That doesn't happen to you without leaving some appreciation of what it means to have your land taken," she said. "It is important when you are pursuing broad policy goals that you don't forget the people who are sacrificing for those goals."

Harris initially had been against the Plains & Eastern project for this reason, but had come around. The project helped the United States keep up with China on high-voltage direct current technology. And as the exact route of the line came into focus, it was clear that a 720-mile route could be devised that didn't require knocking down a single house. By early 2015, she believed the transmission line was beneficial and could be done in a fair way. But she remained wary.

Skelly's enthusiasm and willingness to work with communities along the Plains & Eastern helped ease her concerns. She also wanted to unleash more renewables, which she believed required building interstate transmission. There was no magical device that could provide inexpensive, reliable, and clean electricity without any sacrifices. After decades working in energy policy, she said, "What you realize is that there is no form of energy that doesn't do damage."

While these negotiations were under way, the Energy Department didn't give any indication which way it was leaning. This added to Skelly's rising stress. The Plains & Eastern was well into Skelly's second stage of project development, despair, and headed for the third: the search for the guilty. "Whose stupid idea was this," Skelly said, explaining the retaliatory mood that permeates this stage.

The anxiousness around the office rose as the wait for the federal government dragged on. Cary Kottler, by this time Clean Line's general counsel, sometimes didn't come into the office because he got tired of his co-workers asking him for updates. He kept his telephone close, ready to answer questions from Walsh. "I didn't let a DOE call go to my voicemail," he said. "I once jumped out of a shower to answer a call."

While Energy Department staffers worked on the deal, Tennessee senator Lamar Alexander was putting pressure on Moniz. In

June 2015, he warned Moniz that he had a "serious concern" with the Plains & Eastern project. "Wind is not effective," he wrote in a letter, because it "does not blow when demand is high." A year earlier, TVA CEO Bill Johnson had told Alexander this wasn't true for the Plains & Eastern. Wind from the Oklahoma panhandle, he wrote, "produces at about a 50 percent capacity factor between the hours of 4:00 p.m. and 7:00 p.m., thus contributing to meeting peak demand." But Alexander appeared to be aware that if the TVA added wind, it would either have to burn less coal or shutter one of Alexander's beloved, job-providing nuclear plants.

The TVA was already closing some of its coal plants, as they aged and costs to run them got too high. It was winding down operation at its Widows Creek Fossil Plant in northern Alabama, a giant coal burner that had been commissioned when Harry Truman was president. In September 2015, a train would bring the last few loads of coal. A machine called a dumper would lift each 140-ton car and turn it upside down. The utility planned to demolish the plant's sixty-story reinforced concrete chimney. It had hoped to keep the final unit of the plant open for a few more years, but the cost of expanding the coal ash pond was prohibitive. As the TVA wound down operations, Google announced it wanted to take over the site and turn it into a $600 million data center. The company said it would work with the TVA to secure 100 percent renewable energy to power the facility. Where that renewable energy would come from wasn't specified.

That summer, Skelly had more pressing matters than Lamar Alexander's letters or the glacial pace of federal decision making. After Global Infrastructure Partners had pulled out, he needed to find more money to keep the company going until Clean Line secured a deal.

Michael Zilkha and the Ziffs were not putting more money in.

They had both invested in 2010 and had expected the project to be much nearer completion and a successful exit. National Grid, the U.K. company, had invested nearly three years earlier and didn't want to commit more funds. And with three senators—Alexander, Boozman, and Cotton—speaking out against the project, the Plains & Eastern had a whiff to it that repelled investors.

Getting a new source of funding was critical. The company would soon be spending tens of thousands of dollars a month on lobbyists to defend itself from political attacks. And payroll had grown to nearly one hundred people, a by-product of working on four different lines at once. Skelly could lay off staff to keep the company alive for longer, but at what price? The entire endeavor could appear vulnerable, which would scare off already skittish state regulators.

As the Houston summer set in, the financial situation was getting desperate. Skelly sat down with Cary Kottler, the lawyer, and asked him to prepare a plan if they ran out of cash. He wanted to make sure the company could wind down in a way that didn't leave employees or vendors empty-handed. At regular all-hands meetings held on Fridays, when employees could ask Skelly and the other managers anything, Skelly tried to be honest about the situation without causing panic. "We can promise you an interesting place to work," he told them. "We can't promise you job security."

In the midst of Skelly's search for a new investor, trouble arose in Missouri for the Grain Belt Express Clean Line. Clean Line was asking the state for a utility certificate of need, but it didn't look like Kansas City Power and Light or anything the five-member Missouri Public Service Commission had ever seen before. It was new and was asking regulators to be accommodating. It didn't help that farmers had organized against it and were exerting political pressure. More than seven thousand comments had been submitted to the commission.

At a hearing in June, three of the regulators made it clear they were looking for reasons to reject the request. Commissioner William Kenney explained he would vote no. A former National Football

League quarterback, mostly with the Kansas City Chiefs, he had gone into real estate after retirement and then politics. His rationale for turning down the project was that it didn't serve the state. "I am supposed to look after Missourians and not the rest of the country," he said. "I will vote no because I do not see the benefit to Missourians."

The order, when it was issued a month later, denied Clean Line's application. The commission decided it was more appropriate to consider the impact on Missouri utilities "rather than how it might affect Kansas wind developers or utilities and consumers from other states." Weighing whether the project was needed, the commission asked whether Grain Belt was enough of an improvement to the power grid to justify its $2.2 billion cost. It was an odd question to ask. There was no cost to Missouri customers. Unlike almost every other project that came before the five-member commission, Clean Line was ready to pay its own way. Jimmy Glotfelty said he just wanted to tell state and federal regulators: "We don't need your money. We don't want your money. We just want the pathway."

The Missouri order's logic made Skelly apoplectic; it perplexed Missouri commissioner Robert Kenney. He shared a last name with his fellow commissioner, but little else. Soon after the vote—he voted in favor of giving Grain Belt its certificate—he resigned. On his final day, he issued a scathing dissent that he said grew out of his frustration by his fellow regulators' "narrow and parochial" approach. Missourians bear no risk. They wouldn't bear any cost whatsoever. In denying the request, he wrote, "We are like the Luddites of the nineteenth century, telling the world that we do not embrace new technologies. We are telling the world that we prefer central planning to free markets," he wrote. "I respectfully dissent," he concluded, in his final act as a public official. Then he moved west to San Francisco to work as a vice president for Pacific Gas & Electric. He continued to follow the case.

Kenney said the regulators' logic went too far. "If we had applied that same analysis to the interstate highway system, we would never have had interstate highways built. We are an assemblage of United States, but we are a nation as well," he said. Skelly and Clean Line appealed. But it would take time—and Skelly knew they were running out of money. "Time is the enemy of all deals," he said.

Back in 2009, Skelly had set out to prove that building the infrastructure of a new, cleaner energy system could be profitable. "If this works, it is fantastic," he explained once. "We are proving a new way of doing things. If it doesn't work, I think all these conversations about building a massive grid will become a purely academic concept. Academics will talk about it. Serious people won't." Companies aren't altruistic, and neither are investors. But show them they can make money while investing in cleaner energy, then there would be billions of dollars in pension funds and infrastructure investment funds looking for projects to back. The Obama administration was moving so slowly in handing out its approvals that investors were losing interest; Skelly wanted to demonstrate that doing something good for the environment could also turn a profit.

Skelly finally found a new investor. Bluescape Resources, a private investment fund, said it would invest $17 million in the company, with an option to go up to $50 million and possibly many times more once the Plains & Eastern and other transmission projects moved from development to construction. Skelly talked to employees about how delighted he was to have another serious investor on board that would help them realize their vision.

John Wilder, Bluescape's head and the former chief executive of TXU, a Dallas-based power company, was held in high esteem by the utility world. He would lend his imprimatur to Clean Line. But there was a price to be paid. Skelly got the money he needed to move forward, but the company's valuation was down. If Global Infrastructure Partners hadn't been scared away by the introduction of the Approval Act, it would have been an up round, in the parlance of Silicon Valley. But Wilder had sensed that Clean Line was under stress and had negotiated a 15 percent stake in the company. Clean

Line was still worth $110 million, but that was a little off where it had been months earlier.

Wilder said he was making a bet that the United States needed to get serious about building its energy infrastructure. "China is cleaning our clock," he said, in a conversation about how China was building several high-voltage direct current lines to move enormous amounts of electricity around the country. The United States needed to get serious. It wouldn't be easy. "It will take a lot of courage from public officials," he said. But he was confident that in Skelly, he had found the right person to cajole and convince public officials to get it done.

———

Into the fall, the fate of the Plains & Eastern was uncertain. Skelly and the rest of Clean Line waited on tenterhooks for word from Secretary Moniz. But as they waited, the fate of the project was under assault a few blocks from the Energy Department's hulking building on Capitol Hill.

Congressman Womack had filed his own version of the Approval Act. And after several months, the House bill finally got a hearing. The bill was a sniper's shot aimed at one thing only: the Plains & Eastern. Jared Huffman, a congressman from California who represented a district in California that stretched from the million-dollar bungalows in the North Bay to the marijuana farms in Humboldt County, said as much at the beginning of the hearing: "This bill is aimed at blocking construction of the proposed Plains & Eastern Clean Line project." Over the next hour, no one disagreed.

The testimony offered in that hearing in October that was most compelling and effective came from about as far away from a Bay Area Democrat as was possible in the United States political landscape. It came from a lifelong Republican lawyer who represented an electrical cooperative based in Nacogdoches, in the piney woods of East Texas. And yet despite their geographical and ideological distance, the congressman and the lawyer agreed.

Bill Burchette spoke on behalf of the East Texas Electric Cooperative and three other rural power companies that serve 330,000 Texans in an arc from east of Dallas to north of Houston. He spoke for several minutes without once glancing at his notes.

"I wasn't a big fan of wind power originally because I thought it was too expensive," he said. But the co-ops had discovered that in the northern part of the service area, buying wind power made lots of sense. "We found that we can use that energy on our system and lower our power cost," he said. But they couldn't get the power to its southern members because there was too much congestion. East Texas Electric Cooperative and his other clients had hoped the Plains & Eastern would bypass this electron traffic jam and enable the rural power companies to buy power out of the TVA on less crowded transmission lines.

Speaking deliberately, he explained that his clients saw the Plains & Eastern as an opportunity. "So when we saw this particular piece of legislation," he said, referring to Womack's bill, "we looked at it and said what this [bill] is going to do is basically kill this project." Not only would the bill hurt the Nacogdoches power cooperative, he said, it was just bad policy. Clean Line had gone out and spent tens of millions of dollars. Now Congress wanted to change the rules at the eleventh hour? It was creating an unstable environment to make investment.

Opponents of the Plains & Eastern saw an attempt to usurp state political power. Transmission siting had been a state function. They argued that Section 1222 of the Energy Policy Act of 2005 was trampling on this traditional role. Jordan Wimpy, a Little Rock lawyer, testified at the hearing that the bill would "restore states' rights and state primacy."

The white-haired Burchette could barely contain his scorn. Joe Barton, the Energy Committee chairman, had overseen passage of the Energy Policy Act. "I don't know of a bigger states' rights man in this Congress," Burchette said. They had spoken after the bill about this particular provision. "He talked to me in terms of this is going to allow us to build some bigger projects without having each individual state negate it," he said.

"We are a utility company. We believe in the same principles of state siting. This is different. This is not your ordinary transmission line," he said. The bill made it out of committee, but Womack struggled to get much support for it. He could never get the bill up for a vote. But that doesn't mean it wasn't a success. Arkansas delegates returned home, boasting of their efforts to defeat the Plains & Eastern.

———————————

The year ended on a high note for Skelly and the rest of the company. They had survived a congressional assassination attempt and were still pressing ahead. At the holiday party, held that year on the first floor of the firehouse, Skelly stood up and recited a poem for the gathered Clean Line workers. It was Rudyard Kipling's "If—" a four-stanza exhortation to live life in an upstanding manner, and to have courage of your convictions. "If you can bear to hear the truth you've spoken, twisted by knaves to make a trap for fools, or watch the things you gave your life to, broken," Skelly read. If you can persevere, "yours is the earth and everything that's in it." After a couple of hours, the Clean Line workers shuffled out in search of a nearby bar to continue the party without any of the management team.

Skelly was content. Clean Line would soon deliver a term sheet to the Tennessee Valley Authority. It was for 1,500 megawatts of wind— about 7.8-million megawatt hours a year. He was offering to sell them power at $39 a megawatt hour. The amount would rise by 2 percent a year for thirty years.

For the TVA's just ended fiscal year, the utility had paid more than $50 for the electricity it purchased either through long-term contracts or on the spot market. Clean Line's new offer was a bargain, and considerably less costly than the $65 that Skelly had first proposed in 2009.

Not long afterward, Skelly got more good news. The city of Tallahassee said it wanted to buy power from the Plains & Eastern. It was only 50 megawatts, but it was a customer. And Skelly was negotiating with more. A small utility in Tampa was interested, and there were some talks with the giant Duke Energy utility in North Carolina. Skelly still needed the TVA as his anchor customer, but the other pieces were coming together.

In Missouri, Skelly improvised a different approach to win the first customer for the Grain Belt: he tried to give away electricity. At a lunch in Columbia with a group of municipal power companies, he blurted out an offer to deliver power to them for free. They would have to pay the wind generator in Kansas, but the charge for using the wires would be reduced to next to nothing. "There's a joke in the office," Skelly said later. "If you want a shitty deal to get even worse, send Skelly." But it wasn't all that crazy. They could buy the wind at a low price and get it delivered for next to nothing. "That saves ratepayers money," Skelly said, and at the next regulators meeting "they show up and testify for us." Clean Line would have allies in the state capitol who stood to save millions of dollar a year.

Skelly was willing to offer TVA a deal also. He modified the term sheet, shortening the duration to a twenty-year contract and allowing TVA to buy 500 or 1,000 megawatts. It rushed the update to get ahead of a February meeting of the TVA board of directors.

It is unclear if the Plains & Eastern term sheet was shared with the board of directors. "At this time, TVA is assisting Clean Line with the interconnection process," TVA staff wrote in a memo for the board. Then Dave Berry came up with a creative idea to drop the price even further. In March, Clean Line sent over an idea that it called a "proposed transmission capacity sharing arrangement." It was not an official term sheet, but a two-page explanation of an idea that Berry had devised.

TVA would buy 1,000 megawatts at $29.50 for a megawatt hour. To make this work, TVA would allow Clean Line to move 1,500 megawatts across its system, wheeling power from near Memphis to where the TVA connects with its neighbors. Clean Line could sell this power to Southern in Georgia or Duke in North Carolina. Since the Oklahoma wind farm could generate and Plains & Eastern could move 4,000 megawatts, that left 1,500 megawatts unaccounted for. Clean Line would try to sell those also to TVA's neighbors. The utility would be paid up to $26 million a year for wheeling the power across its system.

It was a complicated deal, but the bottom line was Clean Line was offering to sell TVA electricity at a competitive price. That year, when TVA generated a megawatt hour of electricity from coal, it burned up $28.40 worth of coal from Illinois, Wyoming, or Appalachia. When it used natural gas, it burned $32.50 worth of the fuel to generate a megawatt. That was only the fuel. Operating and maintenance costs needed to be added. By comparison, Skelly's $29.50 a megawatt looked good.

A couple days after the new pitch was delivered, Energy Department lawyer Sam Walsh called Cary Kottler to make sure Clean Line's legal documents were up to date. The Clean Line lawyer knew better than to press him. Walsh had been by-the-book and as close-to-the-vest as possible. But Kottler thought it was a good sign. (Energy Secretary Ernest Moniz had made his decision a month earlier, but no one had breathed a word.)

On the morning of Good Friday, a Clean Line employee in Little Rock heard that a decision would soon be issued. The Energy Department was calling elected officials in Arkansas, and rumors were circulating that an announcement was imminent.

Normally, the office would be empty on Good Friday. But a skeleton crew were at the office to be ready just in case the decision was made public. Kottler was there. Walsh had been pushing him to the limit with requests all year. He was worn out trying to comply. Mario Hurtado was there, as were Jayshree Desai and Dave Berry. They huddled in the small conference room. Skelly was missing. He was on a family ski trip.

No one could get any work done. Hurtado felt a nervous anticipation that reminded him of waiting for a college acceptance letter. Everyone was anxious. If the federal government had decided not to participate, the Plains & Eastern would have no path across Arkansas. After six years, the project would be dead—and the other projects would be on life support. "It was one of those moments of truth," Hurtado said. "All of your efforts are crystallized into one decision."

Normally, Sarah Bray and her public relations team would have had two press releases ready. One if the decision went against the project and another if the decision was good. This time, the management team had decided not to ready a "bad" statement. It was too important. Kottler didn't want to jinx it. He demanded, only half joking, that he didn't want to see any piece of paper with a rejection on it. Anyway, if the federal government said no, not having a press release was the least of their concerns. The company's survival would be a more pressing matter.

If the government said yes, Skelly's quote was ready: "The Department of Energy's decision shows that great things are happening in America today." Then it went on to talk about investing in the grid, jobs, and low-cost power.

Late in the morning, Bray's phone rang. It was her counterpart at the Department of Energy. The decision was about to be posted on the Energy Department's website, he said, but he wouldn't say what the decision was. Bray was so excited she didn't ask which of the Energy Department's websites, or what the decision was. Jason Thomas, Clean Line's vice president in charge of environmental permitting, opened his laptop. He checked the department's Plains & Eastern project website. The Office of Energy Delivery's website. The National Environmental Policy Act website. The press office website. Then he would start again and refresh each one, looking to see if something new was posted.

Nearly simultaneously, Thomas found the decision, and Sam Walsh called Kottler to inform him about the decision. It was a yes. The federal government would go ahead and partner, allowing the project to move ahead. But there wasn't any immediate celebration. "I don't recall celebrating much," Thomas said. "I put my head down and dove into the legal document." So did Kottler. He disappeared into one of the few rooms in the office with a door to read it carefully. The topline decision was good: the government was ready to partner with Clean Line to build the Plains & Eastern. But Kottler wanted to see if there were any unexpected conditions. Was there anything hidden in the document that would snatch defeat from the jaws of victory?

When Kottler emerged, he was smiling. Relief washed over the assembled Clean Line workers. All their effort was not in vain. Kottler's wife and daughters came to the office and he hugged them. His young daughter got bored and tried to amuse herself spinning around on a chair. No one else was bored. Every newspaper and television station in Arkansas was calling, as were national newspapers. The Clean Line cofounders, Skelly, Hurtado, and Glotfelty, made a round of calls to thank supporters.

The Energy Department had concluded that a project to deliver enough energy for 1.5 million homes was worth their backing. "Moving remote and plentiful power to areas where electricity is in high demand is essential for building the grid of the future," it said.

While Clean Line would get its eminent domain, several conditions were placed upon it. Its land agents would need to make at least three attempts to personally meet with every impacted landowner. It had to begin construction by the end of 2018. And the final route chosen by the Energy Department didn't require a single person along the more than 720-mile route to be displaced from their home. It didn't require anyone to buy the energy. It noted that taxpayers would pay nothing. Clean Line agreed to contribute 2 percent of revenue to the government to help fix aging dams. "There is no 'impact-free' routing choice for a large transmission line," the government concluded. Building new infrastructure has costs. But waiting for a perfect solution in which no one is affected means waiting forever. The Plains & Eastern had its approval to build and a government partner that would allow it to cross Arkansas.

15

An Extremely Compelling Price

N ot long after the federal government agreed to partner with Clean Line, the company stepped up its efforts to sign leases with landowners. It was an encouraging time for Michael Skelly and the rest of the company. The Plains & Eastern had transformed from an idea into a full-fledged route that began in Oklahoma's parched appendage and traveled all the way to the humid doorstep of the Mississippi River Delta. But before Clean Line could connect those distant spots on the map with a continuous steel-reinforced aluminum cable, it had to obtain right-of-way easements for every parcel of land along the 720-mile route. Each required a signed legal document. It was an enormous undertaking and one of Skelly's main jobs was to inspire Clean Line's sixty employees and encourage its supporters to capitalize on the project's newfound momentum. "There are a lot of people—allies, environmentalists, businesspeople—who have taken risks and gotten people pissed off at them because they supported us," he said to me. "There is no obligation to win, but there is a very affirmative obligation to work your ass off."

Clean Line employees hung a small whiteboard on a column in the middle of the Houston offices. On it was written how many easements had been acquired. Employees could track the gradual metamorphosis of the Plains & Eastern from something on paper to something that was, if not yet tangible, at least anchored in notarized real estate documents.

One afternoon, slightly after 5:10 p.m., Skelly strolled over to the whiteboard for a twice-daily ritual that he enjoyed. "Can I update it?" he yelled out. It said on the board that 35.7 percent of the required easements had been signed. Standing erect in slacks, a dress shirt, and blue socks, but no tie, he towered over the handful of seated employees working on computers. Several nearby looked up. "For sure," one said.

"What percentage?" Skelly asked loudly, drawing attention to what he was doing. The employee, a young man in his twenties, peered at his computer screen. "36.1 percent," he said, in a tone that suggested he was unsure.

"That's right," said the second employee.

"Tracts?" Skelly asked.

After a beat, the answer came from nearby: "831."

Skelly wiped away the old numbers and wrote the update. He picked up a mallet and hit a small gong on the table under the board.

"Okay, 36.1 percent everybody."

This brief break from the tedium of saving the world one parcel of land at a time ended and several sets of eyes returned to screens. Another three miles of the route had been secured. The Plains & Eastern was that much closer to reality. As Skelly walked away, one of the young workers turned back to his computer and started singing: "Get it on, bang a gong, get it on."

Skelly headed toward the back of the office, turned a corner, and entered a large room. An employee working at a table looked up and said hello. There were rolled maps on another table, piles of papers and three-ring notebooks, but a noticeable lack of art or desks. This end of the building backed up against a parking garage and the cars looked close enough to touch through an opened window.

There were eight large maps, each nearly the size of a bus stop advertisement, tacked up. They covered one windowless wall, turned the corner, and covered part of a second. On the maps, in excruciating detail, was the planned route of the Plains & Eastern, as well as every tract of land and property line. Some had been colored in green. "Green is good," Skelly said. "Green is leased. Paperwork complete." It looked like a giant unfinished coloring book.

Skelly said he expected to be able to come to terms with landowners for 95 percent or 96 percent of the parcels. Eminent domain would be required for the rest. If Skelly had been building a natural gas pipeline, he could have secured federal siting authority and finished this process in a matter of months. But the government had never provided a similar blueprint for transmission lines that crossed from state to state, at least not before Section 1222 of the Energy Policy Act of 2005. Skelly and Clean Line were the guinea pigs.

Two weeks after the Energy Department decided, after years of deliberation, to participate in the project, Arkansas senator John Boozman made a last-ditch attempt to disrupt the bureaucratic machinery. An optometrist and former cattle rancher, Boozman was getting an earful from constituents on the topic of the Plains & Eastern. The senator had tried once to kill the Plains & Eastern through a bill to require what amounted to a state veto over the project. This time he offered an amendment that required the Energy Department to write and issue a report analyzing the line's impact on power prices and then wait ninety days before taking any action. It was clear to anyone paying attention that the amendment would require a standstill that could last the better part of a year.

In the name of fighting climate change, Skelly and the Plains & Eastern were running "roughshod over the state and roughshod over locals," he said later. Boozman was going to make it difficult for Skelly and the rest of Clean Line, or at least make a show that would please vocal constituents such as Julie Morton.

Standing on the floor of the Senate, he promised in a slow, deliberate manner that the "amendment will not cause delays or add additional red tape." New Mexico's junior senator, Martin Heinrich, spoke against the amendment. The Plains & Eastern meant billions in private sector investment and employment for six thousand workers. What the Arkansas senator was offering was a "job-killing amendment," he said. Boozman was defeated by a 55–42 vote.

Skelly felt like a fireman running around putting out conflagrations such as the Boozman amendment on a near-daily basis. Success felt like a moving target. Every time it seemed that Clean Line was making progress, something came up. As the tally on the whiteboard slowly rose, more and more of Skelly's time became consumed with working out a deal with the Tennessee Valley Authority. It was the last major obstacle to overcome before construction could begin.

In June 2016, Clean Line and Invenergy submitted another proposal to the TVA. This time, Berkshire Hathaway Energy, part of the conglomerate owned by Warren Buffett, joined as a partner. Having one of the bluest of blue-chip companies on board was intended to send a clear message to TVA. This was a serious endeavor undertaken by companies that could deliver.

The proposal presented TVA with what it called a "once in a lifetime opportunity . . . to participate in one of America's most significant infrastructure projects of the century and the country's largest renewable energy project." It introduced a snazzy new name for the project: States Edge Renewable Energy Center. (The center of the proposed wind farm was less than fifty miles from Texas, Kansas, Colorado, and New Mexico.)

Up to this point, TVA hadn't said no to the project, but it hadn't said yes either. To make its latest proposal attractive, Invenergy and Clean Line offered an array of options. Would TVA like a twenty-year contract? Perhaps twenty-five? Or thirty-five? Did it want to purchase 500 megawatts, 1,000, or more? And why stop at wind? If the TVA wanted solar, the Oklahoma panhandle offered some of the best solar insolation in the country. States Edge could be both wind and solar power, shipped together on the Plains & Eastern to Memphis.

"We would say, Would you like it with some renewable energy credits? Would you like it with some transmission revenues? Would you like it with some solar on the side? Would you like it this way? That way?" said Skelly.

Attached to the commercial terms, Invenergy and Clean Line filled an appendix with letters of support from large corporations interested in expanding their operations in the TVA service area if they could purchase inexpensive renewable energy. Owens Corning, a giant manufacturer of insulation and roofing shingles, praised the States Edge plan. So did Mars, General Motors, Kellogg's, Ikea, and Facebook. Honda argued that TVA could play a role as a conduit for renewable energy to surrounding utilities. Honda factories turned out Odyssey minivans in Lincoln, Alabama, and lawnmowers and snow blowers in Swepsonville, North Carolina, and the company was interested in signing deals to purchase States Edge electricity energy that passed through TVA's wires. The letters and proposal were sent to Bill Johnson and executives at TVA, and filed away. They were never shown to board members.

Skelly was beginning to wonder if TVA had just been stringing him along for years, filling him with empty hope for reasons he couldn't fathom. The first time he met Johnson, in 2013, they had clicked. They both read thick historical biographies and enjoyed discussing what books they had just finished or planned to pick up. Johnson told him not to bother wearing a tie to meetings. Skelly was happy to oblige. The two men could talk big-picture as well as drill down into the details. Johnson always said he was keen to do a deal, someday and at some future low price.

And it wasn't just Johnson who expressed interest. Clean Line officials met their TVA counterparts on a regular basis. On such a big complicated deal, this was how the power industry worked. The utility industry doesn't make snap decisions. Clean Line was offering something new and different. They expected TVA to take its time and ask a lot of questions. Clean Line was not worried about all the questions. TVA gave every indication they were heading toward a deal.

When Skelly or other wind developers met with Johnson and his staff, they came away feeling positive. Aaron Zubaty, whose company MAP Royalty leased thousands of acres in the Oklahoma panhandle to build a wind farm and hoped to use the Plains & Eastern to carry the electricity eastward, said that every time he left Johnson's office, he always thought a deal was imminent. "We got every feeling that they wanted to do it," he said. "If we had been going to them and over and over they had been saying, guys don't waste your time, we wouldn't have kept going back."

———

Invenergy was a serious wind developer with a track record, just the kind of company that would impress TVA. Invenergy and Clean Line had similar cultures. Anyone who visited Clean Line in Houston and Invenergy's Chicago offices would notice similarities. Both looked and felt more like tech start-ups than traditional energy companies, with young employees in jeans and headphones.

Invenergy was founded and owned by Michael Polsky. He had emigrated from Ukraine in 1976 with his wife, Maya, a smattering of English, a single suitcase, $500, and a mechanical engineering degree. By the time the couple divorced three decades later, the court awarded her an equal share of their $184 million marital estate. It was believed to be, at the time, the largest divorce verdict ever.

Starting out in the wind business in the early 2000s, Polsky had an experience similar to Skelly's with Blue Canyon. Years later, he recounted the story on a podcast in heavily accented English. He developed a wind farm and tried to sell the output to Oklahoma Gas & Electric. He could deliver power at less than 3 cents per kilowatt hour, significantly less than what OG&E paid for the natural gas it burned. But if the utility entered into a deal with Polsky, it wouldn't get any profits. It might be good for customers, but not for the company. The utility turned him away. "They said 'Why would we buy from you and make no money? We'd rather run our own plants and make money that way,'" he remembered being told.

Invenergy was one of three wind companies with large land positions in the Oklahoma panhandle. Along with Invenergy, there was Zubaty's MAP Royalty, which had leased enough acreages to build a 3,000-megawatt wind farm, and Novus Windpower, the company founded by Jay Lobit, whose childhood memory of being blown across the street in downtown Guymon had brought him home to build wind farms. The companies were competing with each other, but also working toward the same goal. If TVA signed up with Invenergy, that would provide the impetus for Plains & Eastern to move ahead. Then Novus Windpower could strike a deal with Duke in North Carolina, or MAP could agree to sell 750 megawatts to Georgia Power or Tampa Electric Co. or even directly to Honda. Helping Skelly sell the line opened up opportunities for all of them.

But Polsky wanted Invenergy to be the lead developer and work on a joint proposal to TVA with Clean Line. Over several phone calls, Polsky prodded, according to Skelly: "We must work together. You must work directly with us. Why would you work with anyone else?" Polsky tried to do a deal with Clean Line to lock up space on the Plains & Eastern ahead of Novus, MAP, or anyone else. Skelly and others on Clean Line's financial team worried that would give Invenergy too much leverage and allow it to squeeze them. They didn't take the deal.

Still, Polsky and Skelly worked together to pitch the TVA on the idea of fuel substitution. Electricity was electricity whether it came from coal or wind. Their electricity was inexpensive, allowed the TVA to avoid fluctuating commodity prices, and had no lingering environmental liabilities such as leaking coal ash ponds. "It is an unbelievable deal," Polsky said.

Making deals was what got Polsky up in the morning and drove him to keep working. "To go out and do the deals, negotiate the deals, to put deals together, that is what I love to do most," he said one time. States Edge was a colossal deal, the biggest, by far, that Polsky had ever worked on. Invenergy's largest wind farm had 156 turbines; States Edge would have 800—or possibly more.

Polsky stood to make a lot of money, as did Skelly, if they could convince TVA and its fellow utilities in the Southeast to sign a deal. But they also believed their wind-plus-wires project would bring less expensive power into the Southeast. During the Great Depression, that was what the TVA had been founded to do. "Electricity for All" had been its motto.

In June 2016, around the time that Clean Line, Invenergy, and Berkshire Hathaway presented their proposal to TVA, Polsky contacted American Electric Power. Based in Columbus, Ohio, it owned Public Service Company of Oklahoma and a dozen other regional utilities from Texas to West Virginia. He talked to them about the Oklahoma wind farm he was developing and tried to interest them in buying some of its output.

The idea intrigued AEP, but they weren't interested in delivering electricity to Tennessee. Why go all that distance when it owned utilities in Oklahoma, Arkansas, Louisiana, and Texas? Why not use some of that bargain-basement electricity for its customers closer to the Oklahoma panhandle? AEP and Invenergy started swapping ideas. The details of their discussions were confidential. Skelly wouldn't find out about them for more than a year.

By the late summer, Jordan Wimpy, the Little Rock lawyer who had testified before Congress on behalf of the Approval Act, filed a federal lawsuit to stop the project. He claimed the Energy Department's decision to partner with the Plains & Eastern was trampling on states' rights. It was a "stunning example of federal outreach," he wrote in the complaint. Skelly tried to keep Clean Line employees from getting discouraged. "I believe that when projects go through shit storms like this," he said, "they emerge as better projects."

Wimpy's dislike of the project was both philosophical and personal. The son of a rice farmer, he grew up in Poinsett County, Arkansas, close to the Mississippi River. His father bulldozed the land into large bays, which he flooded with a few inches of water to grow rice. Poinsett is the top rice-producing county in Arkansas, the top rice-producing state in the country. Poinsett rice farmers didn't like the Plains & Eastern at all. Rice farmers in the county use crop dusters to spread herbicides and fungicides on their bays. Jordan Wimpy's father, Tom, said he worried the Plains & Eastern would interfere with aerial spraying. "They said, 'we'll make you whole.' They don't have a clue what whole is," Tom Wimpy said.

Jordan Wimpy represented two groups of plaintiffs. One was made up of Poinsett County rice farmers. The other was an organization called Golden Bridge created by Julie Morton, Alison Millsaps, and others. The defendant was the U.S. Department of Energy. Clean Line's lawyer, Cary Kottler, filed a motion to intervene to make sure Clean Line's perspective was included. The case was assigned to a federal judge in Jonesboro, Arkansas, about twenty miles north of the two-story granite-and-concrete Poinsett County courthouse. Wimpy would have the home-court advantage.

At the same time, across the Mississippi River in Tennessee, Bill Johnson was in no hurry. Earlier in the year, TVA had put out a call for renewable energy and gotten several responses, including the States Edge proposal. Then he put them in a file and . . . did nothing. Johnson told a reporter from the *Chattanooga Times Free Press* that the

TVA would move ahead with a deal "if it makes sense under our time-table, not someone else's timetable."

As the summer of 2016 turned into fall, it became clear that the TVA would not make any big decisions until after the November presidential election. Skelly's productivity slowed in the final weeks of the campaign. For everyone in the office and many people across the country, it was hard to turn away from the unfolding political soap opera. Nonetheless, in the final few days before the election, General Electric announced it was joining the Plains & Eastern team. It would build three converter stations to change alternating current into direct current, and back again. It was about a $1 billion order, said Skelly, and created an ally of another big blue-chip company. The deal completed the selection of major contractors for the project. Sediver, a French company, would build the zinc-and-glass insulators in Arkansas, and Pelco Structural would make the poles in a factory outside Tulsa. The more than seven hundred miles of power cable would be made in a facility in Malvern, Arkansas, near where the "Electric Queens," wearing crowns of bulbs, paraded in 1914 to advertise the arrival of electricity. Without this $130 million order, the Malvern plant's owners had said they would close the facility and lay off 130 workers.

A couple of days later, Donald Trump won the election to become the forty-fifth president of the United States. Among the thousands of federal appointments he would get to make would be two new TVA board members. Over the previous few months, Trump had made it clear he didn't like renewable energy. He had spoken often on the campaign trail about restoring coal jobs and boosting oil production. When he talked about wind power in the past, it was to rail against turbines planned for Aberdeen Bay near his Scottish golf course. He once sent a letter to the first minister of Scotland to express his displeasure. "Wind power does not work. Don't destroy your coastlines and your countryside with these monstrous turbines," he wrote. In a separate letter, he urged the minister to "stop your mad march into oblivion with these very expensive and highly inefficient (and extremely ugly) industrial turbines."

A few minutes after midnight on election night, as the tabulations came in from Pennsylvania, Wisconsin, and Michigan, Skelly sent an email to everyone at Clean Line. He was still reeling from the election results. But he felt the need to give the people who worked for him some hope that their work was not going to be in vain.

"Needless to say, these election results create additional uncertainty about our business," he wrote. "It's our job collectively to manage through these changes. Our projects address some major challenges that will be there irrespective of who is in the White House. Climate change is real. Our outdated infrastructure is a huge problem for the country. And as Trump campaigned, we have big issues around creating high-quality jobs. Our projects address all of these issues."

Sometimes politicians surprise you, he said, reminding his colleagues that everyone thought the wind industry would sputter when George W. Bush was elected president. The opposite had happened. He reminded Clean Line employees that Duke Energy was coming to the office the next day. Regardless of the election, the North Carolina company was still facing state mandates to add renewable energy and wanted to talk about purchasing wind power off the Plains & Eastern. Everything had changed, but nothing had changed. He signed off, promising to talk more with the staff after the Duke meeting. And then he went to sleep.

Skelly wasn't sure if the election would spell the end of his nearly decade-long effort or give it a boost. The president-elect clearly relished building things, and what real estate developer was going to get hung up on eminent domain? "Trump is talking about private infrastructure, paid for by private investors," he said. "This fits perfectly." Skelly was ecstatic in mid-December when a list of fifty top-priority infrastructure projects had leaked out of the incoming administration. The Plains & Eastern was #9. It was described as a "national security project that can add resiliency to our electrical grid."

But days later he felt a little lost. "I don't know what Trump wants," he said. Then he found a thread of hope and described a scenario in which Trump would criticize the Obama administration for not being able to get the project done quickly and for the delays that nearly killed Clean Line. Then Trump would swoop in and get it done, taking credit for the largest renewable energy project ever. If Nixon can go to China, then Trump can be a renewable champion.

In mid-December 2016, Clean Line employees gathered at the former Firehouse No. 2, Skelly's house, for the annual holiday party. Skelly and his wife were in the process of creating a small compound. They had purchased six nearby dilapidated Victorian houses, jacked them up onto flatbed trucks and relocated them behind the firehouse. They were being rehabilitated and filled with friends and acquaintances. The *Houston Chronicle* dubbed it "Skellyville." The firehouse filled with other residents of Skelly's world. They came from the Midwest and Oklahoma, everywhere Clean Line had employees trying to find pathways through political and regulatory mazes.

As usual, Skelly spoke about the five stages of project development. This time, the fifth stage had a new twist. The final stage was "the most important one of all, which we are trying valiantly to reach: riches and glory for the uninvolved. Trump comes in and says I made this shit happen." There was some laughter. Who cares who takes the credit? Who cares if Trump wants to cut the ribbon and preen for the cameras? As long as it meant that Plains & Eastern got built.

Even though he had been gloomy earlier in the day, that night he was ebullient. He ended by reminding everyone, maybe even himself, why they kept moving ahead, through the moments of euphoria and long days of despair. The Plains & Eastern won't solve global warming, he said. But it is a serious, large-scale project that will have a global impact.

What he didn't say, didn't need to say or want to say, was that Clean Line was in an increasingly difficult position. Too many more delays and the project's economics would begin to crumble. The first cracks had already appeared and been spackled. If contracts weren't signed soon, the cracks would return and grow wider and longer until the project would be condemned.

———————————

In the eight years since work had begun on the Plains & Eastern, the cost of delivering wind power had continued to fall. This allowed Skelly to give Bill Johnson what he said he wanted: a lower price for Plains & Eastern power. Manufacturers were building taller turbines and using material science to construct longer blades. In this case, size matters. Clean Line had first considered a 72-meter-blade turbine, but now was looking at newly available blades that were more than 100 meters long. The extra meters made a huge difference. The three turning blades formed a large circle and drew their energy from the wind passing through that circle. A well-known mathematical equation explains why longer blades helped so much. If the blade length was the radius of the circle, then the area of the circle was πR^2. As the blade length grew, the potential power of the turbine increased exponentially.

The larger turbines caught the more consistent wind a little higher off the ground and could also operate for more hours of each day. The gearbox, induction generators, and other parts that converted the wind into electricity were also improving. In 2009, Skelly had talked to the TVA about charging $70 per megawatt hour. By 2014, it was $40. The States Edge proposal that Clean Line, Invenergy, and Berkshire Hathaway had delivered in 2016 was now offering electricity for about $26.

That was a very good price. Around the globe, an auction in Morocco had produced a below $30 price for wind energy—and that was hailed as a breakthrough. A few months later, the winning bid to build a wind farm in Alberta was $28.

The Clean Line offer of $26 per megawatt hour was also competitive with prices at new gas plants. In early 2017, TVA opened a state-of-the-art $1 billion Paradise Fossil Plant in Muhlenberg County,

Kentucky, running General Electric's newest, most efficient gas turbines. That year, it burned through 27.3 billion cubic feet of natural gas, according to federal data, and generated 3,830 gigawatt hours of electricity. In other words, it used about $20 worth of natural gas for every megawatt hour of electricity. That didn't include debt payments on the $1 billion price tag or labor or maintenance. Of course, it could also be turned on as needed, something you couldn't always do with wind.

Still, TVA passed on Clean Line's latest proposal. So Dave Berry, the head of strategy, came up with a creative solution that would allow a further price cut. TVA would purchase up to 900 megawatts of States Edge electricity delivered on the Plains & Eastern. That left an additional 2,600 megawatts that would be delivered into the TVA region, then transmitted through the region on TVA lines into Georgia, Florida, North Carolina, and Virginia. TVA would be paid for this point-to-point transmission service, and would redirect some of these funds back to Invenergy and Clean Line. (After all, if Plains & Eastern wasn't built, TVA wouldn't get any additional revenue.) The States Edge partners would use these so-called wheeling revenues as a subsidy that allowed them to drop their price per megawatt hour even further. It was a complex deal, but TVA would not assume any risk. If the extra transmission service revenue didn't materialize, they weren't on the hook and still got the agreed-on price.

Berry briefed TVA on the idea. They asked him to update the States Edge proposal in time for an upcoming board meeting. Invenergy and Clean Line sent the new terms on February 2, 2017, accompanied by a letter from Skelly and Polsky. The companies were offering to sell TVA electricity at the "extremely compelling price" of $18.50 per megawatt hour in the first year. Each subsequent year, the price would rise 2.5 percent.

When I mentioned this price to executives and experts in the electricity world, more than one gasped. A price that low was practically unheard of in the power industry. Later in the year, Mexico held an auction and secured 1,300 megawatts for about $20.57 per megawatt hour. The industry press hailed it as record setting.

At this price, the economics of fuel substitution became forceful. Think about it this way. The TVA burned $27.10 worth of coal to generate a megawatt hour. A 900-megawatt coal plant, running at full capacity for six hours, would burn $146,340 worth of coal. Using Oklahoma wind for that same electricity would cost less than $100,000 for a $46,000 savings. For six hours' worth of electricity.

On cold winter days, natural gas can spike upward of $200 per megawatt hour in the region, the TVA told me. Running a 900-megawatt gas plant for six hours at that price would burn more than $1 million of gas. If the Oklahoma wind was blowing those six hours, the same number of megawatt hours via the Plains & Eastern would cost less than $100,000.

Skelly believed these prices would force the TVA to act, and that with the TVA on board, he could secure deals with other utilities. Skelly and Polsky urged TVA to move swiftly, because they needed to start building soon if they wanted to take advantage of federal tax credits set to expire in 2020.

The proposal accompanying their letter said the deal would advance TVA's three pillars: "providing affordable electric power, stewarding the environment and catalyzing economic development." It was an extension cord that would connect the utility—with no congestion and unpredictable congestion charges—to a world-class wind resource. The Oklahoma panhandle also had some of the best solar in the country. If TVA wanted, the proposal again offered to build solar.

The project was shovel ready. It had the permits. Clean Line's right-of-way agents had been very active since mid-2016 and had acquired leases for half of the 720-mile route, without using eminent domain. It would be a $9.5 billion private investment, generating thousands of manufacturing and construction jobs and using enough steel for four aircraft carriers. All that was needed was for the TVA board to approve it. Skelly and Polsky also threw in one more inducement. Invenergy would tear down the turbines outside Knoxville that upset Lamar Alexander so much.

TVA officials asked that Clean Line and Invenergy send a representative to the scheduled meeting on February 16 of the TVA's board of directors in Gatlinburg, Tennessee. Berry, the proposal's mastermind, flew there and sat in a Starbucks across the street from the Gatlinburg Convention Center, waiting for a call that he was needed inside. His cell phone never rang.

A day earlier, the board had gathered behind closed doors. Clean Line assumed they would discuss the new, lower-priced offer. But TVA management didn't give them a copy of the thirty-seven-page States Edge proposal, or the cover letter from Skelly and Polsky that states in underlined boldface that the price was $18.50. TVA management never asked the board to consider it. The only written note presented to the board read: "TVA currently does not have any plans to purchase power from Clean Line."

Richard Howorth, the former mayor of Oxford, Mississippi, and a TVA board member, said he doesn't recall being told the price. All he remembered was hearing that TVA didn't want the electricity it was offered. "TVA did not need the power in the time frame Clean Line was proposing and did not want to obligate ratepayers with a twenty-year contract," he said later.

Bill Johnson had a little different explanation. He said the TVA ran an analysis on the States Edge proposal. Under statute, the TVA can only pursue a new source of power if it is the lowest-cost option. In this case, he wasn't sure it lowered prices. "That is a pretty low price in the wind field, but that wasn't the entire story," he said. "That price was conditioned on revenue from a tremendous amount of wheeling." It would have been easy enough for TVA analysts to discount this revenue. As for States Edge's promise to take all of this risk, he said, "we never believed that."

The analysis also showed that sometimes overnight when power demand is low, enough wind would be coming into the system that

TVA would have to turn down coal plants and nuclear plants, something which could cause operational problems. Clean Line had run its own analysis and disagreed. "We only dipped into their coal," said Skelly. And he said Clean Line offered to stop sending unwanted power for a few hours several times a year if it helped TVA run its grid more efficiently.

Johnson said its analysis showed that Clean Line didn't save TVA any money over the twenty-year life of the contract. "I am both confident and comfortable we came to the right conclusion here," he said. Of course, projecting out two decades takes a lot of assumptions and educated guesses. When I asked to see the analysis, he said that was impossible. I asked about what assumptions TVA made about future gas or coal prices. This too was a secret. I asked how many hours the nuclear or coal plants would need to be turned down to accommodate wind power. That too was off limits. The analysis was never shared with the board of directors—and there is no regulator that scrutinizes the TVA's decisions. Essentially, what he said was you just have to trust us. TVA's chief financial officer, John M. Thomas, who ran the analysis, said that the States Edge proposal nearly passed muster. "It was close," he said. "This was close."

There was no time to figure out what had happened or why the TVA had been presented with record-low electricity prices and shrugged. Skelly was growing impatient and frustrated.

Then the next month, in March, Arkansas's two senators reintroduced the Approval Act. In a letter to the new energy secretary, Rick Perry, they wrote that using the federal government to build transmission lines over the objection of governors was "antithetical to your distinguished record as a champion for states' rights in the face of federal overreach." Skelly was furious when he heard about it. "They filed the bill again to screw us," he said.

A few minutes later, he grew less animated. "At a time when we want private capital to invest in America and create jobs in America, this is the vision that Congress is providing us?" he said. "We spend eight years developing this project and it is attacked by the Congress."

He paused. Clean Line's power lines would create jobs and build infrastructure. "The only jobs this creates is jobs for lobbyists," he said. Clean Line would need to bulk up on Washington, D.C., lobbyists to fight the reincarnation of this bill. That would cost money the company didn't have. And it was paying for lawyers to fight the lawsuit in federal court in Jonesboro, Arkansas, as well as regulatory proceedings in Missouri, Illinois, and Iowa. But there was enough money for one last Hail Mary. If he couldn't get through to the TVA, maybe he could reach their customers.

In May, more than seven hundred executives and their spouses were expected in Savannah, Georgia, for an annual meeting of the Tennessee Valley Public Power Association. From the giant Memphis Light, Gas & Water to the tiny Tishomingo County Electric Power Association, these were the local power companies that bought TVA's megawatts and resold them to homeowners and local companies. They would gather to swap stories and learn about the latest in the utility industry.

The highlight of the weekend was the golf tournament. Clean Line decided to underwrite the golf balls. This year, when the general manager of the Athens (Tennessee) Utilities Board and the CEO of the City of Athens (Alabama) Electric Department hit the links together, they would tee off with balls with "Plains & Eastern" written in orange letters on one side and "2¢" on the other. This referred to 2 cents per kilowatt hour, which was equivalent to $20 per megawatt hour. The small, dimpled messengers of low prices worked great as golf balls, but had no discernible impact on the TVA.

16

Wind Catcher

I t was hard for Michael Skelly to accept or understand. How could the States Edge partners offer such a spectacularly low price for clean, affordable electricity and then get turned away? It didn't make sense. Clean Line had run its own analysis that showed TVA would get a good deal. He couldn't understand why TVA wasn't moving ahead with the proposal.

He had never expected the Plains & Eastern to stall when it was so tantalizingly close to being built. The permits were secured and the federal government had signed on as a partner. Developers in the Oklahoma panhandle were ready to pour concrete bases to erect the first turbines of the largest wind farm in the United States. Negotiations, creative financial deals, and technology had combined to drive down the price of delivering electricity to Memphis. But TVA wasn't budging. The utility's response was to tell Skelly to come back in a few years.

Skelly didn't have a few years. "We are seriously under the gun," he said one afternoon in the Houston offices, scribbling on whiteboard with a red marker. Delay too long and Clean Line would run out of money. Wait even a year and the project would begin losing federal tax credits. But getting the TVA to act was just . . . he shook his head and his voice trailed off. He capped the marker. The Plains & Eastern was stuck in purgatory and he couldn't figure out why or how to get out.

On March 19, 2017, Skelly and Michael Polsky went to the White House to meet with several Trump aides who worked on infrastructure issues. In a memo sent ahead of the meeting, they suggested the TVA was making decisions "based not on the economics of the proposal that has been presented, but rather on the politics." If the new administration could remove political impediments, the jobs and investment would spring forth.

We have spent more than $200 million to design and permit this project, Polsky and Skelly said. The Plains & Eastern would be a one-hundred-year asset, connecting markets and allowing energy to flow across the United States. By our estimates, they said, the line could generate $700 million in savings to consumers. Every year. It is shovel ready. It won't take a cent of taxpayer money. All we need is for the TVA to commit to buying 500 to 900 megawatts of electricity at a rock-bottom price. Invenergy and Clean Line were ready to start spending $9.5 billion to build a giant wind farm and transmission line, creating more than 2,500 jobs in construction and maintenance. D. J. Gribbin, a special assistant to President Trump on infrastructure, listened attentively.

We know it is not your job to call the TVA and tell them to sign a deal with States Edge, Skelly said, but it's a really big project and it could really help a lot of people.

How long have you been working on the Plains & Eastern transmission line? Gribbin remembered asking, somewhat incredulously.

Eight years, Skelly said. A lot of people think we're crazy.

Gribbin didn't make any promises, but said his job was to fix the bureaucracy so that it didn't take this long.

That would have helped, Skelly thought to himself, six years ago when Plains & Eastern's application sat at the Department of Energy. It was water under a crumbling bridge.

————————————

Three days later, on March 22, 2017, politics interfered again.

Lamar Alexander walked to the lectern in the U.S. Senate's windowless chamber. Wearing a red tie and charcoal suit, he took off his glasses and placed them on a side table. Then he read a nine-page speech that was a full-throated denunciation of the Plains & Eastern project and what he called its "giant, unsightly transmission towers."

The Tennessee senator's twelve-minute speech wasn't soaring oratory, but it sent a plainspoken message. If the TVA chose to enter into a contract, Alexander intended to make sure the U.S. Senate would aggressively exercise its oversight duty. This meant he would make sure that TVA's finances were dissected by federal auditors and Bill Johnson would be on regular flights to Washington, D.C., to explain himself. (Bill Johnson had reason to avoid this kind of scrutiny. The Southern Alliance for Clean Energy, a nonprofit that regularly scrutinized and criticized the TVA, would disclose several months later that the TVA had purchased a pair of $10 million corporate jets and a $7 million luxury helicopter, which were used to ferry its executives and board members around.)

Alexander's speech was full of incorrect and misleading statements. He criticized the line for carrying "expensive and unreliable wind." He leveled that charge four separate times, but it doesn't stand up to scrutiny. Just six weeks before his speech, the States Edge partnership had offered TVA power starting at $18.50 and rising over twenty years to a little less than $30. Expensive? The federal government reported that the cost of operating a nuclear power plant in 2016 was between $25 and $26 per megawatt hour. A gas plant was a bit more.

To underscore how unreliable wind was, Alexander talked about the Buffalo Mountain wind farm built in 2001 outside Knoxville. "It is generous to say that it has been a failure," he said. The wind farm had a capacity to generate 27 megawatts, but due to poor winds produced an average of only 4.3 megawatts. "Just 16 percent," he said.

Watching a video of the speech afterward, Skelly fumed: "You know what? That brick phone I bought from Motorola in 1999, it didn't work very well either, but now I have an iPhone." Clean Line meteorologists estimated the wind farms that fed the Plains & Eastern would generate power about 53 percent of the time. In 2016, the nearest working wind farm to States Edge generated electricity 49 percent of the time.

Alexander also contrasted Buffalo Mountain's woeful performance to Watts Bar Unit 2, the nuclear power generating facility recently finished after a multidecade hiatus. The brand-new unit would operate more than 90 percent of the time for the next forty to eighty years, Alexander said, generating emission-free power. It would produce, he said, "no sulfur. No nitrogen. No mercury. No carbon." This kind of tribute was not a surprise to anyone who paid attention to Alexander. Throughout his career, the senator had been a cheerleader for nuclear power.

What Alexander didn't say, and perhaps didn't know, was that Watts Bar Unit 2 wasn't operating. A few days before his speech, a crew building a scaffold inside the plant had accidentally hit a trip button for a large condensate pump. Within fourteen minutes, operators had taken the plant offline and 1,046 megawatts had vanished from the grid.

As Alexander gave his speech, the TVA was working to restart the reactor. A few hours after the senator praised it, a condenser imploded. This time, the plant was partially damaged. Operators had to power the unit back down. TVA later told the Nuclear Regulatory Commission a design error caused the problem. The reactor stayed offline through the summer.

A few hours after the speech, Polsky fired off an email to Gribbin and other White House officials. "If there was any doubt about political pressures keeping TVA from making a business-oriented decision," he wrote, "Sen. Alexander's remarks from the Senate floor today should put them to rest." He never heard back from the White House.

Polsky wasn't the only one upset. Terry Roland, a Shelby County commissioner in Memphis, and a Republican, was so angered by Alexander's speech that he fired back in a press release a day after the speech. "The information he provided, when fact-checked, proves to be outdated and misleading. Shelby County customers along with everyone on the TVA grid will save money by allowing cheap wind energy to be delivered from Oklahoma."

Roland's counterpunch was ineffective. Alexander's deep antipathy to wind turbines had prevailed. He had delivered a clear message to Bill Johnson and the TVA board members.

Five days after the speech, Skelly had dinner with TVA's CEO Bill Johnson. One of Clean Line's main investors, John Wilder, the head of Bluescape Resources, joined them. Skelly and Wilder probed for an explanation and a way forward. Johnson wouldn't budge.

We don't *need* the power right now, Johnson said, maybe check back in a few years. Skelly sputtered. People in the Tennessee Valley were turning on the lights, he said. You *need* power. The question is one of price. At what price do you not burn coal or gas? Skelly said that at $18.50 per megawatt hour, the Plains & Eastern was easily hitting that number. In a few years, there wouldn't be a Clean Line to build a dedicated line to zap megawatts from the windiest and sunniest part of the country. Skelly said he considered offering him power for free, just to see how Johnson would respond. "We thought we could pay them and they wouldn't do it. Clearly price wasn't an issue," Skelly said.

Skelly was deflated. Even after weeks had passed, he still wasn't his usual voluble self. He and other managers were discussing letting employees go. It would be his first layoffs at Clean Line, or Horizon Wind. He was hoping some would leave for graduate school or other jobs, helping him avoid having to tell someone they didn't have a job anymore. They had slowed the leasing effort to conserve cash. Skelly wanted to stretch the company's remaining funds for as long as possible, all the while trying to bushwhack through the politics and regulatory thickets and maybe catch a break. But more time required more layoffs. It was an uncomfortable tradeoff. Still, he held out hope the Grain Belt line could overcome an adverse court ruling. And the Western Spirit Clean Line, the company's New Mexico project, which had long been a low priority, was progressing.

As for the flagship Plains & Eastern, he felt beat down. "You could have Robert Moses come back from the dead and he wouldn't be able to do shit," Skelly said, referring to the legendary New York City power broker who ruthlessly used political pressure and aggressively deployed eminent domain to build many of the city's roads, parks, and bridges."

As the summer heat crept back into Houston and the Gulf of Mexico's humidity marched inland, bringing afternoon thunderstorms, Skelly started hearing rumors. Friends in the industry's supply chain passed along snippets of intelligence. Invenergy was working on something big. So was American Electric Power.

Skelly called up Polsky a few weeks after their Washington trip. What are you working on? he asked.

I can't tell, Polsky said, and got off the phone.

Polsky had been putting together a project that, in broad strokes, would have sounded familiar to Skelly: a large wind farm in the Oklahoma panhandle with a dedicated transmission line moving the power eastward.

Invenergy planned to build a wind farm on the land cobbled together by Carroll Beaman. By 2017, he had signed up enough of his neighbors that his company, CimTexCo, had rights to develop turbines on more than 300,000 acres in the Oklahoma panhandle. The land controlled by Beaman was larger than Oklahoma City, itself a sprawling metropolis that covered more square miles than Los Angeles, Houston, or Phoenix.

Beaman never had any intention of building a wind farm. He planned to sell the land leases to an experienced company. After building up CimTexCo's acreage, he had called Skelly and asked for suggestions of wind developers to contact. Skelly had given him a few companies and numbers. Invenergy was at the top of his list. When Beaman called them, they were receptive. He gave them a price and, to his surprise, they accepted it without countering.

That was in March 2016. Months later, clerks in the Cimarron and Texas counties courthouses began transferring the rights to erect wind turbines on land belonging to old panhandle families—the Dixons, Hanes, Spradlins, Stewarts, Webbs, and others—from CimTexCo to Invenergy. Invenergy acquired leases for about two-thirds of CimTexCo's land.

In June 2016, nearly a year before Skelly called Polsky to ask about the rumors, Invenergy contacted American Electric Power. The utility, once a giant coal burner that had been switching to gas and renewables for several years, wanted to learn more. "I was surprised to find that one developer had so much property," AEP executive Mark McCullough later said. He liked the potential size of the project and how consistently the wind blew across the flat prairie.

While these talks continued, Invenergy held in August what it called the "States Edge Summit" in Chicago to pitch the Plains & Eastern. It invited executives from AEP, as well as Duke, Southern, Dominion, and other utility companies that served customers in the southeastern United States.

"We discussed, I think, the concept in general. There were several utilities in the room. The proposal was 'hey, we're open for business and we are happy to move forward with any of you in any different way that suits you,'" recalled attendee Antonio Smyth, an AEP executive.

By this time, McCullough had put together a team at AEP, including Smyth, to run numbers on States Edge. It didn't take long for the team to see the benefits of a wind farm in the panhandle. AEP started thinking about fuel substitution—the same idea that Polsky and Skelly were pushing. They could use the inexpensive wind when it was available to avoid burning expensive coal and gas.

AEP got "pretty excited about the potential," Smyth said. But AEP had no interest in building a direct current line to send power skimming along an elevated line all the way across the Mississippi River. It owned a utility in eastern Oklahoma and another in western Arkansas, and it wanted to serve its customers in Tulsa and Texarkana. To get the power there, it would build its own dedicated transmission line and get reimbursed by ratepayers for it, plus a nice profit for AEP and its shareholders. It had no interest in the Plains & Eastern.

Over the next months, the outlines of a deal came together. In November 2016, a couple of weeks after the presidential election, AEP and Invenergy executed a joint development agreement. The proposed wind farm would be 2,000 megawatts, half as large as Skelly's envisioned 4,000 megawatts, but still capable of producing as much power as the five largest existing wind farms in the United States combined.

After signing, Polsky continued meeting with Skelly and developing the States Edge proposal. Together their companies built financial models and constructed elaborate pitches. They sent TVA the $18.50 per megawatt proposal. They went to the White House together to talk with D. J. Gribbin.

Skelly flew to the Oklahoma panhandle in late spring 2017 to attend the annual Pioneer Days Rodeo and reassure local officials the project was still viable. With TVA's intransigence and rumors of Polsky's secret dealings, he wasn't so sure anymore, but he wasn't prepared to let it go. Amid the barrel racing, steer roping, and crowning of a rodeo queen, Skelly bumped into the local wind developer Jay Lobit. Skelly told him that things weren't looking good, but he was still working hard at it and didn't have anything else to do. "Don't worry, I will never give up," Skelly said.

By then, the rumor mill was filling in some details. There was an AEP-Invenergy deal in the works. And by the size of the orders that the supply chain was hearing, it was going to be a huge wind farm. Skelly called up AEP executive vice president Lisa Barton and asked her if she was working on a panhandle wind deal. He remembered her telling him that she could neither confirm nor deny.

One rumor that Skelly heard was that AEP wanted to build a wind farm and a dedicated transmission line from the panhandle to Tulsa. He thought this might be a chance to reanimate the Plains & Eastern. Clean Line had a planned and permitted transmission corridor that was 60 percent leased and passed thirty-five miles south of Tulsa. Plains & Eastern could build its direct current line all the way to Memphis with a stop south of Tulsa to drop off 2,000 megawatts. Because of all the permitting and land work done already, Skelly figured this would be AEP's fastest and least expensive option available.

Right before the July 4th holiday weekend in 2017, Clean Line sent a note to AEP asking to enter into formal talks. AEP could use the Plains & Eastern route to Tulsa and Skelly could keep trying to work on extending it through Arkansas and on to Memphis. AEP said it wasn't interested.

On July 26, AEP came out of stealth mode and confirmed everything Skelly had been hearing. It was buying Invenergy's panhandle wind farm, already in the early stages of development, and would build a 350-mile dedicated transmission line to Tulsa. AEP called it the Wind Catcher Energy Connection, a $4.5 billion megaproject. The wind on the Great Plains would provide savings to AEP's customers of more than $7 billion, net of cost, over twenty-five years, the company claimed. A spokeswoman said it was the lowest cost source of power available.

Around the Clean Line offices, people were stunned. They had worked for years to prove that it made sense to build a giant wind farm and move the power east and AEP had stepped in at the last minute and stolen the idea.

In a private moment, Clean Line cofounder Mario Hurtado asked Skelly if it pissed him off. "It doesn't piss me off that much. I don't know why. Maybe it should," Skelly said. He was numb and exhausted. Eight years earlier, the two men had met for Thai food and had begun talking about building a transmission line to enable massive renewable energy development. The idea had seemed outlandish at the time. Now it was becoming real, but someone else was going to build it. "You suffer a lot of indignities in this business," Skelly said. "Maybe someone copying your idea isn't the worst of the indignities."

A day after it unveiled Wind Catcher, AEP held a conference call with investors. "It just looks like a great project," said AEP's chairman and chief executive Nick Akins. One Wall Street analyst asked whether the region needed the power. "I mean, are you shutting other plants? Are you—is demand growing?" she asked.

No, responded Akins, but that wasn't the point. All existing power plants would remain open. Wind Catcher would add a huge amount of inexpensive energy that could offset higher-priced power. "Low energy, very low energy pricing . . . means economic growth," Akins said. AEP knew what the power price would be from Wind Catcher for the next twenty-five years. Amid all the uncertainty—gas price spikes and the like—it was locking in a new source of inexpensive power for customers.

An hour after the call ended, Akins sent an email to some of AEP's top executives, highlighting the importance of the project: "All of you probably know by now, our earnings call went very well and I believe only because of the Wind Catcher project. Almost all of the questions were about that and it certainly carried the day." The project was part of a new AEP, and they needed to deliver the project on cost and on time. "In your hands are the reputation and emerging brand of AEP and I have no doubt you working together as a team can make this happen," he wrote.

AEP filed paperwork in Oklahoma, Arkansas, Louisiana, and Texas—the four states where customers would pay for the new transmission line in their monthly bills. The argument that AEP makes in thousands of pages of filings in those states is similar to Clean Line's proposals to the TVA. Demand for power is flat, but the projected savings from wind were so great that it made sense to accelerate plans to add renewables. Customers want more renewables and being able to increase the amount of available wind and solar power would make the region more attractive to businesses committed to running off green power. The argument was positively Skellyish.

At the bottom of the eight-page Arkansas application was a familiar name. One of the lawyers that AEP had hired to shepherd Wind Catcher through the regulatory process was Jordan Wimpy, the son of rice farmers who had filed a lawsuit to stop Clean Line.

The news was not much better in Missouri. After having lost a 3–2 vote before the state's regulatory commission in 2015, Clean Line appealed for reconsideration. It signed up the Missouri Joint Municipal Electric Utility Commission to a deal to buy power on the Grain Belt transmission line that would save customers across the state millions every year. In August, the commission voted on the project again and approved it unanimously. But it was a meaningless vote.

In an earlier court case, Missouri judges had ruled that a different transmission line must get consent from every affected county before the state could approve the project. The regulatory commission said it was bound by this precedent and couldn't approve Grain Belt since the project didn't have permission from each of the counties where the line would pass through. The decision sends "a clear message that investors contemplating new infrastructure projects should not come to Missouri," Skelly tweeted.

Skelly was being ground down by the news. He had let some employees go. Even after taking this step, funding was running out and his investors were restless. "They are hanging in there, but they get a little grumpy," he said.

Not long after the Missouri vote, Hurtado testified in Oklahoma that Wind Catcher was wasting time and money on developing a new route. "We have spent more than eight years focused on a very similar goal," Hurtado said. The Plains & Eastern was ready to go and could begin construction in 2018. It would cost less and face far fewer uncertainties.

Oklahoma regulators could require AEP to use the Plains & Eastern route if they believed it was better for the state's electricity customers. Some companies, filing briefs in a regulatory case that soon grew to thousands of pages, said AEP was moving too quickly. It had pushed through a giant, no-bid project. And by building 2,000 megawatts of power and putting it into Tulsa, they were making it very difficult for anyone else to develop a wind farm in the panhandle.

Clean Line could have engaged in a regulatory brawl to force this issue. But Skelly wasn't sure if his investors or employees had the stomach for another fight. Then, in late August, NextEra Energy, a Florida utility and power developer, approached him about buying the Oklahoma portion of the Plains & Eastern. NextEra had plunged into renewable development in 2002, about the time that Zilkha Renewable was beginning to grow. By 2017, it had become the largest wind and solar operator in the world. The Plains & Eastern, or what remained of it, would be in capable hands. Skelly signed a letter of intent and began to negotiate a sale.

There were now two giant electric power companies jockeying to build a wind farm in the Oklahoma panhandle and a new path to carry the power eastward. It wasn't exactly what Skelly envisioned eight years earlier in the garage apartment behind his house, but it was something.

This deal marked the beginning of the end for the company Skelly had helped build. It would be stripped for parts like a car in a junkyard. His vision of direct current power lines zipping inexpensive electricity around had faltered. He had wanted to move power seven hundred miles from the dusty ranches of Steinbeck's Joad family to Mark Twain's watery world of Huck and Jim. Instead Wind Catcher proposed moving power half as far, from one end of Oklahoma to the other. The patchwork of alternating current grids serving regional utilities would have to do, and it would mean fewer and higher-cost renewables.

"It was not a giant step forward," Skelly said, "but it was a step forward."

"A partial victory. Can you live with that?" I asked.

"Do I have a choice?"

———

Skelly's wife once told me that he "always believes in and looks for the good in people. Sometimes it is a blinder." Skelly failed to anticipate Michael Polsky working on a deal with AEP that would leave Plains & Eastern stranded. And he believed Bill Johnson when the utility executive told him, again and again, that he was interested in a deal.

Skelly was also undone by his own vision for renewable energy. He was one of a handful of people in the wind industry who had a clear vision of capital budgets in the billions of dollars. He wanted to unleash the power of capitalism to fund an energy transition.

Along the way, the wind power had become a big industry. AEP wanted to build a $4.5 billion project, which was big enough to move the needle for even a giant energy company. The same potential for billion-dollar profits attracted NextEra, an even bigger energy company. Wind was no longer a niche, cottage industry. "It was huge money," said Aaron Zubaty, of MAP Royalty. "This is about Nick Akins going to Wall Street every quarter and talking about a wind project. Once it got to this point, the rules changed." Skelly didn't realize how much the rules had changed.

A generation earlier, the wind industry had met for its annual meeting at a seedy hotel in San Francisco with water stains on the carpet. There had been eight vendor exhibits on tabletops. In 2016, when the industry met in New Orleans, the convention sprawled across several football fields. There were luxurious booths with leased Barcelona chairs and gourmet coffee. Europeans with stylish glasses and expensive business attire had replaced the men wearing socks and sandals.

On December 1, 2017, Jimmy Glotfelty sent a note to the remaining employees of the company he had helped create, telling them he was leaving.

"I have gone thru a rollercoaster of emotions and now am totally at peace," he said. He listed his reasons for being grateful. The Plains & Eastern was fully permitted and had a strong buyer. They had taken on a hard challenge and succeeded. And they had built a company that helped shape many careers positively. He signed off "until the next rodeo."

Two and a half weeks later, Clean Line withdrew its application to inject power onto the TVA grid near Memphis. Three days later, it signed a deal to sell the Oklahoma portion of the Plains & Eastern to NextEra.

Mario Hurtado said the feeling around the office reminded him of watching a relative in declining health. You know they are close to death and make your peace. But when they die, it still hurts.

At the annual holiday party, the remaining employees and several alumni came to the office to drink and swap stories. "We were feeling bittersweet," remembered Cary Kottler, the general counsel. Skelly tried to put on a good face. When it came time for his speech,

Skelly talked about the five stages of project development. There was the bliss of euphoria. Then came despair as the honeymoon ends and the difficulties ahead become apparent. The third stage is the search for the guilty. The fourth is the punishment of the innocent.

And when he got to the fifth stage, the slightly inebriated crowed raised their plastic cups filled with margaritas and many in the room shouted all at once. "The fifth stage is riches and glory for the uninvolved!" Someone else would have to carry the project over the finish line. Maybe AEP. Maybe NextEra. When the party wound down, many of the younger workers went to a hole-in-the-wall bar called the Big Easy Social and Pleasure Club. Skelly went home.

The next morning, Skelly typed out an email to Bill Johnson. It was short and to the point. "FYI, we sold the Oklahoma portion of our line to NextEra yesterday. We have also pulled out of the TVA queue," he wrote. "Merry Christmas!"

Kottler arrived at work a little groggier than usual, thanks to a late night at the Big Easy. Shortly after lunch, he got a surprise in his email inbox. Judge D. P. Marshall Jr. had handed down his ruling in the lawsuit argued by Jordan Wimpy and brought on behalf of rice farmers and other landowners in Arkansas.

Wimpy had attacked on many fronts. The federal government had overreached its authority and acted capriciously. State permission should be required to build a transmission line. The government wasn't trying to improve the country's power grid, it was just trying to enrich a private company. The entire underlying law was flawed. In short, the Plains & Eastern was trying to bend the federal law to feed its greediness.

Kottler read through the eighteen-page decision, growing more excited by the page. The Department of Energy "acted reasonably and carefully, not arbitrarily and capriciously." It had compiled an administrative record of more than 95,000 pages to decide whether to partner with Clean Line. Requiring Clean Line to get approval from Arkansas would unreasonably give the state a "de facto veto over whether this transmission line gets built."

On each page, the judge's decision got stronger and more sympathetic to Clean Line. The judge rejected the argument that this was a "bridge to nowhere," citing multiple studies that stressed the need for more west-to-east transmission. And wind developers were clamoring to sell their power. Judge Marshall dismissed the case.

Kottler sent an email around to the dwindling number of Clean Line employees. We won the lawsuit, he wrote. It is a definitive, clear decision. "Congratulations everyone," he wrote. "This decision holds the door open for future infrastructure projects." Plains & Eastern wouldn't be using Section 1222 of the federal law, but the decision allowed others to follow their path.

A few weeks later, Skelly withdrew the application to partner with the federal government to build the line. They had worked on it for several years, but it didn't matter anymore. Jayshree Desai signed the document for Clean Line.

Arkansas senator Tom Cotton called Julie Morton to tell her the news and offer congratulations. "I can't believe a bunch of grassroots people beat those billionaires," she said.

"You have a lot of persistence," he told her.

"I've been told that."

Over the next few months, the process of stripping Clean Line for parts continued. It sold its wind farm and Western Spirit, the transmission line under development in New Mexico, to another renewable energy company. Rock Island was dead. With two of its projects sold and a third defunct, that left Grain Belt. It was eventually sold to Invenergy.

Some of the core development team, people such as Jayshree Desai and Dave Berry, stayed together and were backed by investors to build more renewable energy and power storage projects. Cary

Kottler went to work for the company that bought Western Spirit. Jimmy Glotfelty took a job with the largest builder in North America of transmission lines. Mario Hurtado created an energy consulting firm. Dozens of other former Clean Line employees are working in the renewable energy industry.

Skelly wasn't sure what he would do next, but was visiting Mexico, where a large renewable and transmission boom was beginning. In the months after the company folded, he talked to various people about opportunities. "I am feeling a little heartbroken," he said.

17

The Network

On a cold, rainy day in October 1954, President Dwight Eisen-hower flew into Detroit. As he landed and exited the plane, the sun came out and the weather cleared. His motorcade drove downtown, passing under overpasses crowded with well-wishers. As he approached Cadillac Square to give a speech, the former general took off his hat and waved it. "We are pushing ahead with a great road program," he told the crowd, "a road program that will take this nation out of its antiquated shackles of secondary roads all over this country and give us the types of highways that we need." It was his vision for something brand-new: an interstate highway system.

If one of the thousands of people who listened to Eisenhower had left the speech, bought a Cadillac Fleetwood, and headed west toward Chicago, the experience would be very different than what it is today. The driver would have idled at stoplights, slowed down while passing through countless downtowns, and been caught behind slow trucks on undivided two-lane roads. Patience and accordion maps would have been twin necessities of such a road trip.

Two years after the speech, Eisenhower signed the National Interstate and Defense Highways Act of 1956. A few years later, motorists heading west from Detroit could enter Interstate 94. The only thing that would slow them down would be stops for gasoline, sleep, and food.

The U.S. power grid remains like the pre–interstate highway system. Electrons, like that imaginary Cadillac Fleetwood, cannot travel far without encountering congestion. It has been almost a century since W. S. Murray, gazing out from the Rocky Mountains, saw the wisdom of stitching together the nation's many grids into one. And while the grid delivers light into every corner of the country, the electrons often must take inefficient roads to get there.

As the twenty-first century unfolds, the North American grid—once hailed as one of the greatest achievements of the twentieth century—is beginning to look and feel outdated. It delivers electricity, but is it up to contemporary environmental and energy challenges? What would it look like if the grid could remove its own antiquated shackles? Michael Skelly suggested an answer to that question. Keep the existing alternating current grid, just as we still have two-lane country roads. But overlay high-capacity direct current lines on top to move large amounts of electricity from one region to another.

Operating independently of Skelly, federal scientists also asked similar questions. Would such an overlay be cost effective? Would it be more reliable? Would it enable cleaner power? In January 2016, the scientists published a peer-reviewed paper that modeled a network of high-voltage direct current power lines that could shuttle power across the country. Their answers were resoundingly positive.

Alexander "Sandy" MacDonald, one of the country's top meteorologists and who had run the federal government's Forecast Systems Laboratory for nearly two decades, started work on the paper not long after the United Nations Climate Change Conference in 2009. One day during the conference, under a gunmetal-gray sky, MacDonald grabbed a beer with friends from the American delegation. Sitting in a café next to the harbor in Copenhagen, Denmark, where the conference was taking place, the group talked about the future of energy. The Copenhagen talks don't matter, one of the delegates said. Nothing will change because there are no workable alternatives to the fossil fuels that can power the global economy. Renewables are too expensive and too intermittent. How can you depend on solar when a cloud system puts it out of commission, or wind when sometimes the air is as still as a graveyard?

The discussion sparked an idea for MacDonald. He wasn't a climate scientist or an energy expert. But he understood weather. His friends were right. In any one place, it was sunny some days and windy some days. But the United States was continental. He had a hunch that somewhere there would be good sun and somewhere good wind.

MacDonald started to find a way to test if his hunch was correct. He didn't know much about transmission lines, how they worked or how to design a system to move all this power around. But he had the good fortune to team up with a brainy young mathematician named Christopher Clack, who was fascinated by complex, nonlinear dynamic systems such as power grids. He had received a National Academy of Sciences postdoctoral grant that allowed him to move to the Denver area where he began working with MacDonald.

MacDonald figured it would take six months to do the research. It ended up taking nearly six years. Clack built a computer model called NEWS, the National Energy with Weather System, to run simulation after simulation, looking at weather patterns and how electricity could move across a network of direct current lines. Each simulation required so many calculations that it took four days to run.

The output of these simulations was promising. By 2030, the United States could cut its carbon dioxide emissions by 80 percent, using only existing technologies. And the cost of power wouldn't rise. It would be cheaper. It didn't rely on some hoped-for storage breakthrough. All it required was building a network of Plains & Eastern–like lines.

A grid of new giant power lines would be as transformative as "the transcontinental railroads of the nineteenth century, and the interstate highway system of the twentieth century," Clack and MacDonald concluded in a paper they later published about their research. It would create a national market for electricity, allowing large amounts of wind and solar energy to be built profitably while slashing carbon dioxide output.

MacDonald and Clack weren't the first to suspect that building a network of transmission routes, like the interstate system, would create a more efficient and less expensive grid. But previous efforts had all been thought experiments. In 1976, during a decade of energy crises, a congressional report looked into building a coordinated national grid. The authors concluded the idea might work, but they couldn't tell. "At present there are no computer models sufficient to handle the enormous amount of data and calculating required in such a study," the report noted. Forty years later, computing power had caught up. Clack's model incorporated forty thousand possible locations for renewable energy with large numbers of weather and transmission permutations.

Not long after publishing the paper, Clack left the world of government and academic research. He wanted to work at a faster pace. Unable to take the weather data or computer model with him, he built a new one that he said is even more sophisticated and does a better job of reflecting reality. He calls the new system WIS:dom, for Weather-Informed energy Systems: for design, operations, and markets.

The new model incorporated more years of weather data and finer geographic and temporal detail. He has called it "a planning model on steroids." He is looking at the entire globe, not just North America. And he has incorporated batteries and other systems that can effectively store electricity. "Storage makes a big difference," he said, "but it doesn't change the thesis. It's not 'oh, you don't need transmission.' We found the exact inverse. If you have storage, you want transmission and lots of it, to ship power around. The storage allows you to use the transmission system even more efficiently."

Clack's new computer model has run simulations that are so

encouraging he wants to convince grid operators, politicians, and power companies to get to work on building a new grid as quickly as possible. The United States could eliminate more than 90 percent of its carbon emissions with a network of direct current transmission lines, while also cutting costs by 20 percent.

Clack had followed Michael Skelly and Clean Line's efforts. We should learn from Skelly and keep moving ahead, Clack said. He was in a hurry. There was a big problem that needed to be solved, and the clock was ticking. One lesson, he said, was don't build point-to-point direct current lines like the Plains & Eastern from Oklahoma to Memphis. We need to think about a network of transmission with fewer long lines and more short direct current lines.

With more, shorter lines—like interstate highway segments connecting cities—every state will have an off-ramp to download power. "You can have off-takers in every state, and have jobs in every state and tax revenue in every state," he said. This will ease some of the political problems.

The network will be more resilient and reliable. If a storm knocks out part of a new network, other routes will be available. The power grid can be cleaner, cheaper, and more reliable.

All of this is doable, Clack emphasized. The technology all exists. "All we have to do is learn to have people agree on this stuff," he said. "Essentially we can start building today."

Clack's vision is wonderful, but it's worth remembering that it lives in the binary world of computers. The real world is messier, with politicians and their preferences, outdated statutes, angry residents, entrenched utilities, and a dozen other factors. It's hard to put Julie Morton's passion into a computer model, or Lamar Alexander's dislike of wind turbines.

Maybe Michael Skelly and Clean Line will turn out to have blazed a trail that others can follow. The second mouse gets the cheese, as the saying goes. What is usually left unsaid is that the first mouse gets the trap. That's what happened to Cape Wind. Jim Gordon filed permits for the United States' first offshore wind farm in 2001. Then he ran into a political buzz saw. An array of Nantucket homeowners, including Walter Cronkite, the Kennedy clan, and Lamar Alexander, fought the proposal. His permits were challenged twenty-five separate times. He won all of the appeals, but the fight took so long, Gordon ended up emotionally and financially drained. His original agreement to sell power also expired. Gordon finally gave up in November 2017. Announcing his decision, Gordon said he hoped his legal legwork would "hopefully make it easier for other offshore wind developers that follow."

He may well get his wish. The U.S. offshore wind industry is poised to boom in coming years. New and planned offshore wind farms are farther out to sea than what Cape Wind envisioned, and therefore less visible. More political groundwork has been done to build coalitions and opposition to the plans has eased.

Perhaps the Plains & Eastern will turn out to be the first stirrings of a new power grid. Denver billionaire Philip Anschutz is developing a Wyoming–Southern California line. And there's an effort to build a direct current line from the windy parts of Minnesota into the big eastern grid. It would run underground, along Canadian Pacific Railway tracks. "It is both a market opportunity and a moral imperative to build these lines," said Winslow "Trey" Ward, head of the SOO Green Renewable Rail project.

Even Bill Johnson, the TVA head, sees a future for a more networked grid. "The grid we have today is like the intrastate highway system. Not the interstate highway system," he said. "That is a limitation on how efficiently it can move power from one region to another. . . . I do think you will see, at some point, greater movement of power from region to region."

One day in the fall of 2017, when the Plains & Eastern was a zombie project, dead but still moving, Skelly and I went on a bike ride around Houston. Earlier in the day, Energy Secretary Rick Perry had proposed a rule to prop up nuclear and coal plants. His idea was widely criticized by most of the industry and environmentalists as ham-handed, backward-looking, expensive, and polluting.

We started off at the downtown Clean Line offices. Skelly led the way through several city streets and onto a path that ran westward along the Buffalo Bayou. As we headed away from the skyscrapers, there were places where a thick layer of mud left behind from recent Hurricane Harvey flooding covered the path. There was a tree with a plastic tarp deposited by the floodwaters twenty-five feet above the path.

We stopped on a pedestrian and bike bridge over the bayou and I asked him what the city had been like during the storm. Like what he had seen in a war documentary, he said. The Firehouse hadn't flooded, so he had invited a family taking refuge at the crowded convention center to stay with his family. He urged his friends on social media to do the same.

Skelly knew the city well as a biker. He guided me through neighborhood streets to trailheads that led to paths along bayous. We biked about ten miles in all. Toward the end of the ride, Skelly said he wanted to show me something. We headed to the outskirts of downtown, and he got off his bike next to a low metal fence near the confluence of White Oak Bayou and Buffalo Bayou. The fence was intended to keep people out, but he hopped over and asked me to pass the bikes.

"What I'm going to show you has taken the Parks Board six years," he said, requiring permits from the city, the sewer board, Harris County, and the Army Corps of Engineers. Having cleared the fence, we got back on our bikes and pedaled under a bridge and around a corner. We came on a handsome new public gathering spot, still under construction and not yet open. There were steps leading down to the water where people could eat lunch and watch the two bodies of water come together.

"When this is open, it will open up that part of downtown," he said. The bicycling ballot initiative he had led provided funding to link an assortment of previously unconnected paths. When construction is complete, the paths will link together, allowing riders to go from neighborhood to neighborhood, commute downtown, or just travel throughout the city. Skelly was creating a biking network for Houston.

His work at Clean Line was not all that different. He had tried to create the beginning of a new electrical network that connected far-flung parts of the country. By running a transmission line from Oklahoma to Tennessee, or Kansas to Indiana, he was tethering regions to each other. He wanted to build transmission lines that could stitch the country together with a series of poles and steel-reinforced aluminum cable. It was a tall order.

The first time I met Skelly, he had told me he wanted to fix a balkanized power grid owned by state-based and inward-looking utilities. "They weren't looking at the big national picture," he said. He wanted to fix the grid so that clean, inexpensive power could flow freely to where it was needed. He wanted to build a profitable company, but he also wanted to demonstrate that building a new grid of high-capacity transmission lines could be profitable. If he could, others would invest in what he felt was a critical part of the energy transition. Skelly wasn't successful. The Plains & Eastern remains an idea, not a reality. But Clean Line showed what could be done. It is up to others to walk through that door.

Building an electricity system that is reliable, safe, affordable, and clean is one of the great challenges of this generation. The world is electrifying. The use of air-conditioning is on the rise around the world. Electric cars are now mass-produced. The modern world runs off electricity, yet we have an antiquated mode of moving electricity around.

And we need to build more transmission. If we connect regional markets, we can share inexpensive renewable power in a way that makes the power grid more reliable and cleaner, and keeps costs low.

Electricity and the power grid will be with us for a long time. The sooner we begin to build a new grid to bind us together more effectively, the more of a fighting chance we'll have to slow, and ultimately reverse, the carnage created by changing climate.

"You only get one life, right?" Skelly once said. "You might as well do something that is interesting and is challenging and is exciting. If it weren't all those things, it wouldn't be worthwhile."

Thanks and Acknowledgments

I began work on *Superpower* during a year I spent as an Energy Journalism Fellow at the University of Texas. Thanks to everyone there who provided me with camaraderie and support: Fred Beach, Tom Edgar, Christa Hopkins, Claudia Martinez-Castañón, Joseph Noel, Mauricio Pajón, Gary Rasp, Josh Rhodes, David Spence, Melinda Taylor, and Dave Tuttle—and alphabetically last, but deserving of special thanks: Michael Webber.

I wouldn't have had a fellowship without the backing of Leslie Eaton and Matt Murray at *The Wall Street Journal*. Thank you. At the *Journal*, I also owe a debt of gratitude to Elena Cherney, Jamie Heller, and Karen Pensiero for supporting this project. Thanks to my most understanding editors, Miguel Bustillo and Lynn Cook, and my fellow Texas scribblers, Erin Ailworth, Chris Matthews, and Brad Olson.

Thanks to Merlin Peterson who came in on a day off to unlock the doors of the Pope County Museum in Glenwood, Minnesota. Dustin Rader and James Tew at the High Plains Technology Center in Woodward, Oklahoma, helped me make it to the top of a wind turbine, and then back down again. Thanks for convincing me to get on the whale's tail. It was worth it.

My sister, Rachel, was my first reader and counselor. Having your help meant more than I can express. David McCormick helped me develop a small notion into a full-fledged idea, and Ben Loehnen worked with me until the idea became a book.

Thanks to everyone who agreed to tell me their side of the story. The employees at Clean Line had their hands full running a business without a writer poking around, but they made time for me. I hope the book rewards their openness. Sarah Bray makes a couple of cameos in the book, but that doesn't reflect the amount of time she gave me. And, of course, Michael Skelly was unfailingly generous with his time. Thanks for letting me come along for the ride.

Finally, to Laura and the boys, thanks for being there when I left on this journey and also when I came back.

Sources

This book is a work of investigative reporting. It has been built from hundreds of interviews and thousands of pages of documents. Piecing together this story required spelunking through regulatory dockets that can run thousands of pages to emerge with an insightful comment or illuminating detail. It also meant going through equally voluminous federal documents. The Plains & Eastern draft environmental impact statement ran 1,640 pages; the appendixes were even longer. Some of these documents were readily available. I obtained others through Freedom of Information Act requests. I interviewed as many people as possible, sometimes multiple times for clarification. I relied on as many contemporary emails and memos as possible to double-check recollections. If not otherwise specified, quotes were given to me directly in interviews. Scenes described were either witnessed by me or described to me by multiple people involved. I interviewed more than 150 people for this book. Everyone involved with this story had the opportunity to tell me their side of what happened.

1 – No Man's Land

The National Centers for Environmental Information maintains a weather station at the Guymon, Oklahoma, airport and is the source of local wind speed information. The National Renewable Energy Laboratory maintains maps of solar resources. The "crowbar hole" quote is from John Gunther's *Inside U.S.A.* (revised edition), Harper & Brothers, 1951.

The No Man's Land Museum in Goodwell, Oklahoma, has back issues of the *Guymon Daily Herald*. The museum is also where I learned the odd history of a part of the country that no state wanted to claim.

The best descriptions of life during the Dust Bowl come from Timothy Egan's great book *The Worst Hard Time*, Houghton Mifflin, 2013, and a book-length collection of Caroline Henderson dispatches for *The Atlantic* titled *Letters from the Dust Bowl*, University of Oklahoma Press, 2003.

Several books cover the history of wind energy in the United States. Robert W. Righter's *Wind Energy in America: A History*, University of Oklahoma Press, 2009, and Peter Asmus's *Reaping the Wind: How Mechanical Wizards, Visionaries, and Profiteers Helped Shape Our Energy Future*, Island Press, 2001, are both good overviews. Kate Galbraith and Asher Price's *The Great Texas Wind Rush*, University of Texas Press, 2013, picks up the story as the modern wind industry began to grow.

Christopher F. Jones's *Routes of Power: Energy and Modern America*, Harvard University Press, 2014, is a good history of energy infrastructure.

My source for the sixty-two hogs per person statistic is the 2012 Census of Agriculture, Oklahoma, Vol. 1, issued in May 2014. The local pork-processing facilities are owned and operated by Seaboard Corp. The company's annual report with the Securities & Exchange Commission (SEC) said the Guymon facility slaughtered 20,500 hogs in double shifts.

The federal government maintains detailed records on electricity generation. Most helpful are the Energy Information Administration's EIA-860 and EIA-923 data files. These are the source of the claim that the 4,000-megawatt Oklahoma panhandle wind project would have been one of the largest power projects in the country. The number of megawatts of solar generation installed by SolarCity and Tesla are detailed in SolarCity's SEC filings and quarterly investor letters.

The Intergovernmental Panel on Climate Change's Fifth Assessment Report, published in 2014, concludes that "the globally averaged combined land and ocean surface temperature data as calculated by a linear trend show a warming of 0.85 [0.65 to 1.06] °C over the period 1880 to 2012." This is in the *Climate Change 2014 Synthesis Report*.

2 – *E Pluribus Unum*

Details of Houston's electrical history came from *Origin and History of Houston Lighting & Power Company*, published by the company in 1940. It reproduced a copy of the 1898 *Houston Daily Post* article about the boiler explosion.

A good overview of the North American power grid can be found at https://www.energy.gov/oe/services/electricity-policy-coordination-and-implementation/transmission-planning/recovery-act-0 (accessed July 2018).

The National Academy of Engineering published their list of greatest feats as *A Century of Innovation: Twenty Engineering Achievements That Transformed Our Lives*, Joseph Henry Press, 2003.

Nick Brown's "end of discussion" quote is from a House Energy and Commerce Committee hearing on July 26, 2017.

Details about Joseph Hinds and his printmaking shop are in document 2053, in Vol. 5, and document 2184, in Vol. 6, of *The Papers of Thomas A. Edison*, edited by Paul Israel and published by Johns Hopkins University Press in 2004. Hinds also wrote a remembrance in 1910 titled "In the Beginning." It appeared in the January 1910 issue (Vol. 2, No. 9) of the periodical *Popular Electricity in Plain English*.

The "no deleterious gases" quote is from "Interior of the Laboratory," an article in *Frank Leslie's Illustrated Newspaper* on January 10, 1880. I also consulted Vaclav Smil's *Creating the Twentieth Century: Technical Innovations of 1867–1914 and Their Lasting Impact*, Oxford University Press, 2005, and Paul Israel's *Edison: A Life of Invention*, Wiley & Sons, 1998. The electrification of J. P. Morgan's house is told well by Jill Jonnes in *Empires of Light*, Random House, 2003. Some of the history of Edison's Pearl Street power plant is from "The Birth of an Industry," an article written by John W. Lieb in *Electrical World* (Vol. 88, No. 11) from September 9, 1922. I also relied on the magisterial *Networks of Power: Electrification in Western Society, 1880–1930*, Johns Hopkins University Press, 1983, by Thomas P. Hughes.

Edison's "uncharted sea" quote was from a *New York Times* 1922 article cited in *The Quotable Edison*, University of Florida Press, 2011.

The axe-wielding crew's "crusade against the copper and steel wire octopus" is from a *New York Times* article titled "Wires and Poles Cut Down" from April 7, 1889. Another source is *The Power Brink: Con Ed—A Centennial of Electricity* by Alexander Lurkis. Details of the Frankfurt line is in Hughes's *Networks of Power*. Information from the 1902 census is in Louis C. Hunter and Lynwood Bryant's *A History of Industrial Power in the United States, 1780–1930, Volume 3: The Transmission of Power*, MIT Press, 1991.

Details of Samuel Insull's 1913 speech and his consolidation of electricity systems in Lake County are in *Journal of the Franklin Institute*, Vol. 175, No. 6, from June 1913. His speech is titled "The Production and Distribution of Energy."

Details of the train crash under Park Avenue are in "Official Inquiry into Train Wreck" in the January 10, 1902, edition of *The New York Times*; this is the source of the Seth Low quote. I also consulted the Historic American Engineering Record for "New York, New Haven & Hartford Railroad, Cos Cob Power Plant" in the Library of Congress under HAER No. CT-142-A. Another key source was W. S. Murray's book: *Superpower: Its Genesis and Future*, McGraw Hill, 1925.

Lake Murray in South Carolina is named after him, and Coy Bayne's book *Lake Murray: Legend and Leisure,* Bayne Publishing, 1999, has biographical information, as does his obituary in *The New York Times* on January 10, 1942. The *Chicago Tribune* article "A Revolt Against Coal Tyranny," appeared on September 12, 1921. The "huge trough filled with smoke and steam" quote is from the article "Sky Can Be Cleared" in the May 27, 1894, *Chicago Tribune.* The "smoke is the incense" quote is cited in "Businessmen Against Pollution in Late Nineteenth Century Chicago," by Christine Meisner Rosen in the August 1995 issue of *Business History Review.*

The W. E. Mitchell presentation was at the Pacific Coast Convention of the American Institute of Electrical Engineers, in Pasadena, California, in October 1924. It was titled "Interconnection of Power Systems in the Southeastern States." The quote about the "world's largest integrated, centrally controlled pool of electric power" is from Hughes's *Networks of Power.*

The analysis of the Eastern Interconnection that Aaron Bloom led is "Eastern Renewable Generation Integration Study" and is a technical report issued by the National Renewable Energy Laboratory as NREL/TP-6A20-64472 in August 2016. The RE Futures study that Paul Denholm worked on is the first volume of "Renewable Electricity Futures Study" by Trieu Mai, et al., and was also issued by the government as NREL/TP-6A20-52409-1 in 2012.

The Michael Morris paper is "Building the Next Interstate System" and it appeared in the January 2006 issue of *Public Utilities Fortnightly.*

3 – Braulio Carrillo

Dolores Skelly provided me with the Skelly family history. I fact-checked everything I could. I was unable to find any manifest to prove that Michael Skelly arrived via the SS *America.* I have no reason to doubt it, but was unable to verify it.

For an overview of Donald Perry's work, I relied on his "The Canopy of the Tropical Rain Forest," in the November 1984 issue of *Scientific American*. The "Jacques Cousteau of the tropical forest" quote is from "Visiting the Great Potoo," by Roger B. Swain, in *The New York Times*, October 5, 1986.

The 1987 letter noting the region around Golfito had "very serious economic and social difficulties" was written by George Evans in May 1987. The letter is titled "Description of Peace Corps Volunteer Service" and records Skelly's time in the Corps. I also reviewed several letters Skelly wrote at the time to his family in Virginia.

For information on Costa Rica's fauna, I consulted "Mammals of the La Selva–Braulio Carrillo Complex, Costa Rica" by Robert M. Timm, et al. It appeared in *North American Fauna*, Vol. 75, published in 1989 by the U.S. Fish and Wildlife Service. John Vidal wrote an informative article on Costa Rica and its embrace of ecotourism and renewable energy in "The Figueres Family Led Costa Rica's Revolution, and Now Its Green Revolution." It ran in *The Guardian* on May 14, 2012. I relied on several news articles about the tram, including "Costa Rica Aiming for 'Ecotourism'" in *The Washington Times* (1993) and "Aerial Tram Gives Visitors Toucan's Eye View of Forest" in *The Seattle Times* (1995).

New World Power Corp. filed annual reports to the SEC, which helped me understand its liquidity situation. Information about the Tierras Morenas wind farm came from a General Accounting Office report GAO/RCED-98-154, "Climate Change: Information on the U.S. Initiative on Joint Implementation," issued in June 1998. I also relied on a 1998 Global Power Report article, "Energia Global Pays $25-Million for a 24-MW Wind Project in Costa Rica."

The Kenetech award from *Discover* magazine was in the October 1993 issue. The Gerald Alderson quote is from an interview in *The Wall Street Journal* from December 20, 1993. The *Journal* covered the company's subsequent bankruptcy on May 30, 1996. The company's boast of 5 cents per kilowatt hour, equivalent to $50 per megawatt hour, is from "Power Firm Races into the Wind" by Jonathan Marshall in the August 29, 1994, edition of the *San Francisco Chronicle*. Further information on the company is in Alexis Madrigal's *Powering the Dream: The History and Promise of Green Technology*, Da Capo, 2011.

Galbraith and Price cover the Big Spring turbines as well as Texas's adoption of a renewable electricity requirement in *The Great Texas Wind Rush*, cited above. Information about TU Renew is in the W. R. Poage Legislative Library at Baylor University. Utilities and their interviews with Texans is recounted in "Listening to Customers: How Deliberative Polling Helped Build 1,000 MW of New Renewable Energy Projects in Texas," a June 2003 National Renewable Energy Laboratory report.

The quote "all the electricity presently consumed" is from "Seizing the Moment: Texas at the Energy Crossroads," a brochure from the Texas Renewable Energy Industries Association, found in Baylor University Texas Collection's vertical file on Wind Power. David Sibley's quote "play and stay" is from "Texans Honored for Promoting Wind Energy" in Environment News Service from May 3, 2000.

4 – Johnny Rotten

Information about Florida Power & Light, or FPL Group and now a part of NextEra Energy, is in the company's annual SEC filings.

The Zilkha family's fascinating history is from many sources. *From Baghdad to Boardrooms: My Family's Odyssey* by Ezra Zilkha and Ken Emerson, was self-published in 1999. The *New York Times* obituary of Maurice Zilkha, published in May 28, 1964, provided additional information about the family, as did "The Baby King," a *Time* magazine article from November 13, 1972. Information on ZE Records and Michael Zilkha's marriage to Cristina Monet is from two articles in *New Musical Express*—"Why Is This Man Hip but a Complete Failure?" by Paul Rambali in December 1981 and "Ze Zound of Ze Ze" by

Adrian Thrills in October 1981—and a wonderful article in the April 1984 *New York* magazine "Les Enfants Terribles de Rock 'n' Roll" by Anthony Haden-Guest. This article notes that "Disco Clone" was named by *Melody Maker* as its single of the week. The quote "the most fashionable label in the world" is from *The Face* and was cited in an August 13, 2009, article in the *Daily Telegraph* by Thomas H. Green headlined "Mutant Disco from Planet ZE." The "I don't give a damn" quote is from the December 1981 *New Musical Express* article. The slasher movie financed by Zilkha is *Maniac* and featured the tagline "I warned you not to go out tonight."

Details of the Zilkhas' Houston oil company come from SEC filings as well as the only article I could find where Michael Zilkha was actually quoted. It was "A Wildcatter on the Tame Side" by Allen R. Myerson in the March 20, 1998, *New York Times*. The Sonat proxy statements provided biographical details of the Zilkhas. The loss of 150,000 jobs and general description of Houston in the 1980s oil bust is from Steve Frazier's *Wall Street Journal* article "City Under Stress: Oil Recession Plunges Houston into a State of Mental Depression" from March 26, 1986. The sale of Zilkha Energy to Sonat is detailed in a November 24, 1997, *Wall Street Journal* article. Michael Zilkha filled in the rest of the story in a lengthy interview.

The source of the data about year 2000 energy—450 gallons, 104 nuclear plants, and one billion tons of coal, etc.—is from the federal government's Annual Energy Review 2000 (DOE/EIA-0384) issued in August 2001. This is also the source of the "continued growth and reliance" quote.

Information about Sonny Bono and his plans to dismantle wind turbines is in a December 24, 1989, *Los Angeles Times* article, "A Second Wind for Energy Industry" by Jenifer Warren.

The history of the wind industry in this chapter, including the AD 947 traveler, is from Vaclav Smil's *Energy in World History*, Westview Press, 1994. More information about the 1880s Cleveland wind generator can be found in Righter's *Wind Energy in America*.

Details about Nicholas Humber's memorial is from "Wind Energy Pioneer Gets a Moving Memorial" by Brian O'Neill in the October 25, 2001, *Pittsburgh Post-Gazette*.

5 – East 11th Street

The history of the East 11th Street wind pioneers has never been written at length. Robert Righter's book *Wind Energy in America*, cited above, has some information in chapter 9. Many of the people involved are still alive and agreed to interviews.

Other sources include the following newspaper articles: "11th St. Tenants Tilt with Windmill and Con Edison" by Robert McG. Thomas in the November 13, 1976, *New York Times*; "State Tells Con Ed to Buy 2 Kilowatts—From a Windmill" by Linda Greenhouse in the May 6, 1977, *New York Times* (source of Con Ed's "we'll be there" quote); "This Was Their Quest," a *New York Times* editorial from May 8, 1977; "A Remarkable Story from the Ghettos of New York's Lower East Side," by Stewart Dill McBride, Christian Science Monitor News Service, printed in *Beaver County Times* on November 11, 1977; and "519 East 11th St.: Neighbors Rebuild Hopes" by Joyce Maynard in the July 8, 1976, *New York Times*. This last article is the source of the "junkie or a pimp" quote.

The Energy Task Force wrote up their activities in "Wind Power for City People," May 1977. This is where I found Roberto Nazario's poem. Alva Tabor III's Master of Architecture thesis at the Massachusetts Institute of Technology—"Solar Energy Development: A Self-Reliant Technology in Search of a Self-Reliant Economy" from June 1977—provided additional details.

Con Edison filed three "special permission applications" with state regulators that provide details. They are numbered EL-1656, EL-1659, and EL-1660 and are useful sources of information. More broadly, insights into Con Ed's operations are from Alexander Lurkis's *Power Brink*, cited above. Information on Jacobs wind turbines is from *Chasing the Wind: Inside the Alternative Energy Battle*, by

Rody Johnson, University of Tennessee Press, 2014. The "I'm kind of a freak" quote is from the article "Wind Power History: Marcellus Jacobs Interview," in the November/December 1973 issue of *Mother Earth News*. His cited testimony was at a hearing of the Subcommittee on Energy of the Committee on Science and Astronautics, 93rd Congress, 2nd Session, on May 21, 1974.

I found information about John O'Sullivan's meeting with Bill Clinton in *First in His Class* by David Maraniss, Touchstone, 1995. The Supreme Court case referred to is *American Paper Institute v. AEP Svc. Corp.*, 461 U.S. 402. It was decided on May 16, 1983.

William Heronemus's proposal was written about in an article by John Walcott in *Blair and Ketchum's Country Journal*. It was reprinted in the Congressional Record of October 9, 1975 (94th Congress, Vol. 121, p. 32,947.)

6 – The Battle of Blue Canyon

Most of this chapter comes from interviews with people involved in the development of the Blue Canyon wind farm. Details of the regulatory fight can be found in the extensive record of the "Application of Blue Canyon Windpower II LLC for Establishment of Purchased Power Rates and a Purchase Power Contact with AEP-Public Service Company of Oklahoma Pursuant to PURPA." It is cause PUD 2003-00633 at the Oklahoma Corporation Commission. The testimony of Wayne Walker and Michael Zilkha was particularly informative, as were the original application, briefs, and commission orders.

I found information on the chartering of Burk Royalty Co. in the January 27, 1935, edition of *The Dallas Morning News*.

The estimates of how much it cost Public Service Company of Oklahoma to generate power from the Anadarko plant was based on testimony about heat rates and federal reports of natural gas costs. Information on the closure of two units and operating rate of a third is from federal electricity data. PSO reported that 22 percent of its power came from wind in a fact sheet from a "PSO Facts" publication December 31, 2016. Information about its forecast for 2024 came from figure 32 in its Integrated Resource Plan, dated September 2015.

7 – A Cure for Insomnia

The story of Goldman Sachs's purchase of Zilkha Renewable came from interviews and documents from both parties. Here are additional sources of quotes and facts:

The $380 million wind farm is Maple Ridge near Lowville, New York. The other projects cited here are Pine Tree Wind Farm in California and High Prairie Wind in Minnesota. The "one of the most forgettable tall buildings" quote is from "Shadow Building," an article by Paul Goldberger in the May 17, 2010, *New Yorker*.

The history of using the federal tax code to promote energy is from a Congressional Research Service report titled "Energy Tax Policy: History and Current Issues," by Salvatore Lazzari, updated June 2008. The 1979 calculation that 35 percent of the cost of generating nuclear power was subsidized is from "Nuclear Economics: Taxation, Fuel Cost and Decommissioning," by Duane Chapman. It was a draft report to the California Energy Commission (revised October 1979) published by Cornell University. I also relied on "Federal Tax Subsidies for Electric Utilities: An Energy Policy Perspective," a 1980 *Harvard Environmental Law Review* article by Christopher P. Clay.

The "helped spawn and push" quote on natural gas use of credit is from "How Unconventional Gas Prospers Without Tax Incentives," a December 1995 article in the *Oil & Gas Journal* by Vello A. Kuuskraa and Scott H. Stevens. Another Congressional Research Service report, this one from 2015 and written by Molly F. Sherlock, "The Renewable Electricity Production Tax Credit: In Brief," was a good source.

Details of the planned Aroostook wind development are in the Maine Public Utility Commission docket 2008-256. The "astonishingly audacious" quote is from a joint filing by the Connecticut

Department of Public Utility Control and the Connecticut Office of Consumer Counsel with the Federal Energy Regulatory Commission, docket EL08-77-000.

The *Wall Street Journal* article "EDP Deal Taps U.S. Energy Market" by Erica Herrero-Martinez in March 2007 provided details about the purchase of Horizon by the Portuguese company.

8 – "Tell Me If I Sound Like a Liberal"

The financial details in this chapter about election fundraising and expenditures come from the Federal Election Commission, especially forms 2 and 3 filed by Skelly's campaign. His television advertisements are all on YouTube. The polling data and analysis was by Greenberg Quinlan Rosner Research. The "draws to a dead heat" quote is from a report by the firm, "Strong Demand for Change in Texas' 7th Congressional District."

The quote "we should appreciate America" is from the March 22, 1983, issue of *The Observer*, the Notre Dame student newspaper.

John Culberson's quote about "a water hose down a gopher hole" is from the *Houston Chronicle* on October 25, 2000. Details on Jim Culberson are from a 1997 obituary in the *Chronicle*. The "sow salt" quote is from a National Public Radio segment by David Welna that aired on April 20, 2001. "Reliably Republican" is from a House Race Hotline article "Ya Mess with the Bull Ya Get the Horns" in April 2008.

Culberson's "subsidized by you and me" quote is from a June 23, 2008, post in the *Houston Chronicle*'s "Texas on the Potomac" blog. The author, Alan Bernstein, covered the race. His coverage was helpful in understanding the dynamics. Also helpful was Stuart Rothenberg's article "How Long Are the House Democrats' Really Long Shots?" in *Roll Call.*

As for Joanne Herring, the best source is George Crile's *Charlie Wilson's War*, Grove Press, 2003. The "Roman orgy" party? *Texas Monthly* wrote about it in "Reversal of Fortune" by Mimi Swartz in

December 2007. If you can find it, read "Roman Rollicking in Texas," in *Life* magazine, Vol. 47, No. 6, from August 10, 1959. Sally Quinn's articles in *The Washington Post* appeared on December 17 and 18, 1978, and were titled "Blond Bombshell from the Texas Boom Town" and "The Glamorous Consul."

Herring told me about her phone conversation with Culberson. I asked close to a dozen times, but Culberson declined my requests to participate in this book. People who worked for Culberson on the campaign provided me with the mailers that linked Skelly with Nancy Pelosi. Culberson's NASA gaffe was written about in the *Houston Chronicle* on July 23, 2008.

Information on the Katy Freeway expansion came from multiple sources, including email correspondence with Raquelle Lewis at the Texas Department of Transportation and a 2008 TxDOT publication "Building a Legacy" produced by Parsons Brinckerhoff. This furnished details on maintenance costs and vehicle count. The "quadruple by-pass" quote is from a *Houston Chronicle* article from January 13, 2008.

Details about condemnation proceedings are in news reports and the January 2014 opinion filed in the Fourteenth Court of Appeals in the case of *Caffe Ribs v. Texas* (No. 14-12-00401-CV).

9 – Clean Line Energy Partners

Details on wind and solar power generation in 2009 are from tables 3.1.A and 3.1.B in the Energy Information Administration's "Electric Power Annual." The *20% Wind Energy by 2030* report was issued by the U.S. Department of Energy in July 2008 and is numbered GO-102008-2567. The "significant amount of new transmission" quote is on page 93.

The "interstate transmission superhighway" quote is from "Wind Energy Bumps into Power Grid's Limits," by Matthew L. Wald in the August 27, 2008, issue of *The New York Times*.

The "holy shit" quote and President Obama's response are from

Jonathan Alter's *The Promise: President Obama, Year One*, Simon & Schuster, 2010. Additional details, including the "moon shot" quote, are in Ron Suskind's *Confidence Men: Wall Street, Washington, and the Education of a President*, HarperCollins, 2011. I supplemented these accounts with my own interviews.

The "Wind Energy Capital" is from Texas Senate Resolution 134 in the 77th Session. Other details about the constraints in and around McCamey, Texas, are from the Texas Public Utility Commission's "Report to the 78th Legislature, Scope of Competition in Electric Markets in Texas." There were quarterly reports filed on the CREZ lines. The final one, No. 17 in December 2014, provides a useful overview as well as details on the size and cost of the program. The "friendly territory" quote is from Rick Perry's remarks on June 2, 2008.

The idea for a "National Electricity Backbone" was integrated into a DOE January 2004 report titled "National Electric Delivery Technologies Roadmap."

Details on T S Ary are from a number of sources: his National Mining Hall of Fame biography; a March 29, 1991, *Washington Post* article, "Mining's Top Prospector Hits a Nerve"; a May 18, 1988, *Washington Post* article, "Three for the Short Term"; and "Mining Chief Assails Species Act and Nature 'Nuts,'" in the March 22, 1991, *Denver Post*.

A July 2009 Clean Line pitch deck provided the "taps into" and other quotes as well as the cost and timetable estimates. The rest of the information in the chapter comes from several interviews with the people involved in early funding for Clean Line.

10 – Euphoria

Much of this chapter is built on repeated interviews with Clean Line employees. The International Energy Agency's "crossroads" and "catastrophic and irreversible damage" quotes are from page 37 of its World Energy Outlook 2008.

The Energy Information Administration's U.S. Energy Mapping

System was helpful in seeing the power grid and its substations, which allowed me to trace Mario Hurtado and Michael Skelly's car trip around Arkansas.

In this chapter and subsequent chapters, I write about the Plains & Eastern project. There is a long regulatory record in Arkansas, Oklahoma, and Tennessee, but the central document is the seven-volume environmental impact statement, issued by the Department of Energy (EIS-0486) in October 2015.

Details of the Kingston coal ash spill come from several sources: the administrative order and consent agreement—docket CER-CLA-04-2009-3766—signed in May 2009; the "Root Cause Analysis of TVA Kingston Dredge Pond Failure on December 22, 2008" by AECOM (Project No. 60095742, issued June 25 2009); a July 2009 TVA presentation by Anda Ray, "Update on Kingston Ash Spill"; an August 22, 2009, article in the *Knoxville News Sentinel*, "TVA Faces More Than $3 Billion in Work"; and a *60 Minutes* piece from October 2009 titled "Coal Ash: 130 Million Tons of Waste."

The 2009 article by German, Swiss, and British scientists that spelled out the idea of a carbon budget appeared in *Nature* on April 30, 2009, titled "Greenhouse-Gas Emission Targets for Limiting Global Warming to 2° C" by Meinshausen, et al. This is the source of the "needed urgently" quote.

President Obama's first weekly radio address was delivered on January 24, 2009, and is the source of the "accelerate the creation" quote. His "inaction and denial" quote is from remarks at the United Nations Climate Change Summit on September 22, 2009. Senator Lamar Alexander's speech in response, the source of the "giant 50-story wind turbines" line, was delivered the following day on the floor of the Senate.

The "without precedent in scale and ambition" and "without a dedicated transmission" quotes are from a May 27, 2009, proposal from what was then called Great Plains to the Tennessee Valley Authority. Skelly wrote a letter to Jim Howell accompanying the proposal, which I also cited. Most of the proposals to TVA and similar documents were obtained through either the Freedom of Information Act or the company itself. The "renewable energy pipelines" quote is from a Clean Line presentation from July 18, 2009.

Details of the Buffalo Mountain wind farms are from federal EIA 860 and EIA 923 data files. Michael Polsky's contention that he lost $6 million is from an article in *Inc.*'s November 2017 issue by David Whitford. The reference to $80 to $85 per megawatt hour prices for TVA's contracted wind comes from an interview with Rob Manning, a former executive vice president of TVA, as well as TVA documents such as the President's Report to the Board of Directors from August 2013. It is notable that in September 2017, the TVA Office of the Inspector General looked at these wind contracts and concluded "TVA's decision to enter into long-term wind power contracts has not proven to be in TVA's economic interest."

The "express train" quote is from page 187 of the large multi-transmission organization study known as the "Joint Coordinated System Plan '08" or the JCSP study.

11 – Chickens and Eggs

Clean Line's application in Arkansas to build a transmission line was filed with the state's Public Service Commission under docket 10-041-U. The order was issued in January 2011. Skelly's quotes are from filed testimony in this case.

The Valerie Boyce quote about the law being "silent" on merchant transmission is from her testimony before the state Public Service Commission in December 2010. This is also the source of Colette Honorable's and Emon Mahony's back-and-forth—and Skelly's "cloud of ambiguity" and "time and money" quotes. A transcript is in the docket cited above. Paul Suskie's quote is from a July 8, 2010, article by Housley Carr in *Electric Power Daily* titled "Entergy, SPP Denied Participation in Ark. Case."

A good overview of the legal history of eminent domain can be found in "The 'Public Uses' of Eminent Domain: History and Policy," by Errol E. Meidinger in the Fall 1980 edition of *Environmental Law*. The courts established the government's power of eminent domain in the 1896 Supreme Court case *United States v. Gettysburg Electric Ry.,* 160 U.S. 668.

The history of Arkansas Power & Light mostly comes from Stephen Wilson's *Harvey Couch: An Entrepreneur Brings Electricity to Arkansas,* August House/Little Rock Publishers, 1986. The Central Arkansas Library System's online *Encyclopedia of Arkansas* provided additional details, including a copy of the "cheapest and best servant" advertisement. It's here: http://www.encyclopediaofarkansas.net/encyclopedia/media-detail.aspx?mediaID=7141 (accessed July 2018). The more hens promotion is from an article by D. Clayton Brown in the *Red River Valley Historical Review* in 1978 titled "Hen Eggs to Kilowatts: Arkansas Rural Electrification." This article is also the source of the percentage of homes that received electrical service.

The Oklahoma application was filed with the state's Corporation Commission and under cause PUD 2010-00075. The federal application was filed in July 2010 and titled "Plains & Eastern Clean Line: Project Proposal for New or Upgraded Transmission Line Projects Under Section 1222 of the Energy Policy Act of 2005." The "vexing challenge" quote is from the federal application as are details about its cost and expected timeline.

The Chinese high-voltage line was written about in a July 9, 2010, release from the Xinhua China Economic Information Service, "World-Advanced Direct Current Transmission Project," and in "Xiangjiaba–Shanghai Transmission Line Switched On" on the China Energy Newswire.

Skelly's letter to Lauren Azar, with the "desire to invest" quote, is from July 11, 2011. Daniel Poneman's letter with the "last resort" quote was sent on April 5, 2012. It is worth noting here that these documents are not, as far as I know, collected in any central depository. All are, or should be, available through FOIA requests. That is how I got most of them.

The story of the Pacific Intertie is fascinating. Here are three broad sources: Joshua D. Binus's 2008 master's thesis at Portland State University, "Bonneville Power Administration and the Creation of the Pacific Intertie, 1958–1964"; the federal Bureau of Reclamation 1997 history by Toni Rae Linenberger titled "The Pacific Northwest–Pacific Southwest Intertie"; and Gene Tollefson's 1987 book, *BPA and the Struggle for Power at Cost* published by the Bonneville Power Administration. The "six San Francisco's" quote is from Binus's thesis.

The "essentially solved" quote is from "Work Done in the Soviet Union on High-Voltage Long-Distance D-C Power Transmission," by A. M. Nekrasov and A. V. Posse in the August 1959 *Transactions of the American Institute of Electrical Engineers*. The "outmatch the Soviet Union" quote is from a January 11, 1963, article by William M. Blair in *The New York Times*. President Kennedy's quote is from his remarks at the Hanford, Washington, Electric Generating Plant on September 26, 1963.

The "ranks with the pyramids" quote appeared in an article from the September 14, 1970, *Los Angeles Times* by George Getze titled "Northwest-Southland Power Tieup to be Dedicated Friday." Floyd Dominy's "This project will make our nation the world leader in an exciting new transmission technique" is from the article "A New Power Giant Materializes on the West Coast," which appeared in the August 1965 edition of *Reclamation Era*. Lyndon Johnson's quote "if we just ignore dissention and distrust" is in Gene Tollefson's 1987 book, cited above.

12 – Julie, Virgil, and Janey

The meeting called by Glenanna O'Mara was covered in the January 11, 2013, edition of the *Southwest Times Record*, a daily paper in Fort Smith, Arkansas. Wanda Freeman wrote the article under the headline: "Cedarville Landowners View Power Line Maps." Descriptions of the routes, including Alternative 4-D, are in the Plains & Eastern's draft environmental impact statement from December 2014.

The spiral-bound book printed by Clean Line was titled *Guidebook for Arkansas Leaders.* I reviewed the November 2013 edition. The "while these payments are not required" quote is from this.

Skelly's quote "they just won't move" appeared in the *Houston Chronicle* on November 12, 2012.

National Grid's investment and quotes are from a November 27, 2012, press release issued by both companies. Skelly's "Hoover Dams" quote is from an article in the *Journal Record* (Oklahoma City) on November 27, 2012.

Julie Morton provided details of her life in a series of interviews. Her house is practically a depository of papers related to the fight against the transmission line. I reviewed the petition cited in the chapter.

Jane Summerson, the government environmental compliance officer who ran the meetings, told me about the chartreuse signs with black lettering. Additional details of the public meetings are from either Summerson or official DOE transcripts. This is where I found Julie Morton's "another American Revolution on your hands" quote. Specifically, see appendix Q, chapter 2, page 1218, of the draft environmental impact statement, cited above.

Michael Brune's quote, "we have to be in support of something," is from an onstage appearance he made at the *Wall Street Journal*'s ECO:nomics conference in April 2016.

The Arkansas Valley Electric Cooperative's solar farm has a capacity of a half megawatt and produces power only a quarter of the time. That's about 1,095 megawatt hours a year. Assuming the Plains & Eastern's capacity was 3,500 megawatts and had a 50 percent capacity factor, that's 15,330,000 megawatt hours a year.

The Minnesota farmers' uprising against the proposed transmission line is worthy of a novel. Until then, the best source is Barry M. Casper and Paul Wellstone's *Powerline: The First Battle of America's Energy War*, University of Massachusetts Press, 1981. I also relied heavily on a Minnesota Public Radio series *Powerline Blues* from 2002. It is archived here: http://news.minnesota.publicradio.org/

features/200212/08_losurem_powerline/ (accessed July 2018). Many of the specific details come from the Minnesota Historical Society's collection, including its oral history of the fight, and documents from the attorney general and Department of Public Safety. This is a rich source of many of the quotes used in this section. This is also the source of the list of equipment and matériel brought to quell the uprising.

The quotes from Phil McMahon are from a videotaping of the symposium I watched at the Pope County Museum.

13 – A Week from Hell

Information about Clean Line's request for information and solicitations for interest by wind developers for both Plains & Eastern and Grain Belt was announced in Clean Line press releases in August 2013, January 2014, and May 2014. The $20 price was in a Clean Line presentation in April 2014. The $70 "all-in cost" was from a May 2009 proposal to the TVA. Skelly's "compelling value proposition" is from a May 23, 2014, article in *Platts Megawatt Daily* by Housley Carr.

Details about TVA's new CEO, Bill Johnson, come from several sources. These include two articles in Raleigh, North Carolina's *News & Observer*: "Ousted Duke Energy Head to Lead the TVA; Ex-Progress CEO Was Fired by Duke in July," by John Murawski on November 5, 2012, and "Heir Apparent Steady in Storms" by Dudley Price on December 21, 2004.

Information about the Duke-Progress merger is from SEC filings, as well as coverage in the *Charlotte Observer* and *Wall Street Journal* on July 11, 2012. The *Journal*'s revealing July 6, 2012, article, "Behind Duke's CEO-for-a-Day—Ex-Chief Gets Job Back Within Hours" by Joann S. Lublin, Dan Fitzpatrick, and Rebecca Smith, was very useful. The "autocratic" quote is from Jim Rogers's July 11, 2012, statement to the North Carolina Utilities Commission, part of docket E-7 Sub 1017. Information on Johnson's TVA salary is in the annual SEC filing from November 2013 and coverage in the *Chattanooga Times Free Press*.

Randy Spicer's quote came from a video of the TVA board meeting archived on the TVA's website. Details about the cost of power generation are from a report to the TVA board on August 22, 2013, and other sources.

The Alexander-Fincher letter is dated May 13, 2014, and Johnson's reply is from June 23. Senator Alexander's quote, "100-yard-tall, monstrous structures," is from a May 20, 2005, *Knoxville News Sentinel* article by Scott Barker. Details about Alexander's Nantucket house are from several sources, including an Associated Press article from June 14, 2005, by Hilary Roxe headlined "Alexander's Financial Disclosure Surprises Wind Advocates." This article is also the source of the "I don't want other people to have to look at them either" quote. I verified his real estate holdings by reviewing U.S. Senate Financial Disclosure Reports and county real estate records.

The "most direct assault" quote is from J. J. Stambaugh's May 24, 2005, article "Senator Wants Moratorium on Windmills" in the *Knoxville News Sentinel*. Alexander introduced Senate Bills 1034 and 1208 in 2005. Alexander's "not our grandmother's windmills" quote is from the May 13 2005, Congressional Record (109th Congress, 1st Session, Vol. 151, No. 63). The "flashing red lights" is from the June 9, 2005, Congressional Record (109th Congress, 1st Session, Vol. 151, No. 76). Neil McBride's recollection of the Neyland Stadium photograph is from an interview.

The "enduring eyesore" quote from Pope County is in an April 2, 2015, email submitted to the DOE and part of the federal environmental impact statement. The first Clean Line–Invenergy proposal was dated January 8, 2015, and was submitted by what was being called the Panhandle Wind Energy Center; I obtained a copy from a FOIA request. The "Bill Johnson is clearly pushing" email is internal Clean Line correspondence.

Senator Boozman's "If a project is not good for Arkansas" quote is from a press release issued concurrent with the filing of the Approval Act in February 2015.

Skelly's "We believe in the merits of the project" quote is from a *Pine Bluff Commercial* article by John Lovett that appeared on February 12, 2015.

Information about the Iowa Utilities Board rulings on Rock Island is in the dockets E-22123 through E-22138. Details on the Smith–Hodges-Copple wedding are from participants, photographs, and the April 2015 edition of the *Chapelwood Chimes*.

14 – This Is Not Your Ordinary Transmission Line

Energy Secretary Ernest Moniz described his thinking to me in an interview, which is the source of the "back to square zero" quote. His recollections are backed up by Department of Energy documents. The "significant congressional interest" quote is from an April 27, 2015, briefing to the secretary, obtained through the FOIA. The Womack "highway for power" quote is from a January 15, 2015, letter to Moniz. The "last resort" quote, cited in an earlier chapter, is from an April 5, 2012, Daniel Poneman letter.

Details on the payments to counties—the $99,000 one-time payment to Van Buren County and $5.7 million over forty years, for instance—come from the March 25, 2016, "Participation Agreement" between the DOE and Clean Line. Schedule 4 has the details.

The Senator Alexander letter expressing "serious concern" was from June 11, 2015. The Johnson-to-Alexander letter that notes a "50 percent capacity factor" is from June 23, 2014.

Details about the closure of the Widows Creek power plant is from "Last Load of Coal Delivered at TVA's Widows Creek Plant," a September 19, 2015, article by Tim Omarzu in the *Chattanooga Times Free Press*. Details on Google's plans for the plant are from http://googleblog.blogspot.com/2015/06/a-power-plant-for-internet-our-newest.html (accessed July 2018) and correspondence with the company.

The Grain Belt case before the Missouri Public Service Commission is docket EA-2014-0207. The order issued on July 1, 2015, notes there were 7,200 public comments. William Kenney's quote "look after Missourians" is from a transcript of the June 2, 2015, hearing. Robert Kenney's dissent was filed on August 7, 2015. His quote "We are an assemblage of United States" is from an interview for this book in 2018.

The Bluescape investment was covered in a press release in July 2015.

Quotes from Jared Huffman, Bill Burchette, and Jordan Wimpy are from the October 28, 2015, legislative hearing. A YouTube video of the hearing is available here: https://youtu.be/4ClzmWWQvYE (accessed July 2018).

The TVA reported in its annual SEC filing for the fiscal year that ended in September 2015 that it spent $950 million on purchased power to obtain 18.837 million megawatt hours, for an average price of $50.43. This filing is also the source of the $28.40 of coal per megawatt hour and $32.50 of natural gas. See the "Fuel Expense per kWh" table on page 19.

"TVA is assisting Clean Line" is from the minutes of the TVA's Disclosure Control Committee meeting on February 16, 2016. Clean Line delivered its second term sheet on January 27, 2016. The "proposed transmission capacity sharing arrangement" was delivered on March 21. I obtained these proposals, or at least redacted versions of them, via FOIA requests to the TVA.

The Department of Energy agreed to participate with Clean Line on the Plains & Eastern on March 25, 2016. It released a press release, a "Participation Agreement," and a "Summary of Findings." All are good sources of information on the project.

15 – An Extremely Compelling Price

The Senator Boozman quote about running "roughshod over the state" is from a 2018 interview for this book. His quote that the "amendment will not cause delays" is from the Congressional Record for April 19, 2016, page S2189.

Invenergy and Berkshire Hathaway Energy submitted a bid proposal to TVA on June 30, 2016, titled "States Edge Renewable Energy Center." This is the source of the "once in a lifetime opportunity" quote.

Details on Michael Polsky come from David Whitford's *Inc.* article, cited in the notes for chapter 10, as well as a June 4, 2007, Associated Press article "Energy Magnate's Wife Stands to Get $184M in One of Nation's Biggest Divorce Awards." I also interviewed Polsky for the book, although the interview was conducted early in my reporting and he then turned down subsequent interview requests. Also instructive is a June 2016 interview he gave on "The Energy Gang" podcast from Greentech Media. His "unbelievable deal" quote is from the podcast. In its June 2016 proposal to the TVA, Invenergy said the largest wind farm it had built was a 511-megawatt facility in Iowa, which had 156 turbines.

Details on Invenergy's dealings with American Electric Power are from sworn testimony in the Oklahoma Corporation Commission's review of the Wind Catcher Energy Connection application, cause PUD 201700267. See, for instance, Novus Windpower LLC's filing on December 19, 2017, and Paul Chodak's testimony from July, September, and especially December 2017.

The lawsuit filed by Jordan Wimpy is *Downwind v. U.S. Department of Energy* in the Eastern District of Arkansas (case 3:16-cv-00207-JLH). Rice production data is from the U.S. Department of Agriculture's National Agricultural Statistics Service. The data I used was from the 2017 survey.

Bill Johnson's quote "if it makes sense under our timetable" is from a September 2, 2016 *Chattanooga Times Free Press* article. The reference to "Skellyville" comes from a Lisa Gray column in the *Houston Chronicle* from February 9, 2014.

The two letters written by Donald J. Trump to Alex Salmond, first minister of Scotland, are dated April 19 and May 2, 2012. Skelly provided me with the email he sent to employees immediately after the November 2016 election.

The Morocco wind prices were reported in the REN21 "Renewables 2018 Global Status Report." The Alberta prices were reported by the Alberta Electric System Operator and provided to me by Tara de Weerd. Details of the Mexico auction can be found in numerous industry press articles. Details on the operating costs of the Paradise plant in Muhlenberg County (yes, it's the subject of the John Prine song) are derived from EIA-860 and EIA-923 data files.

The final Plains & Eastern proposal was delivered on February 2, 2017. A letter from Michael Skelly and Michael Polsky accompanied the proposal. It is the source of the "extremely compelling" quote. The "TVA currently does not have any plans" is from minutes of the January 17, 2017, meeting of the TVA's Disclosure Control Committee. TVA senior vice president Janet Brewer told me in an email this was the only information provided to the board.

Some odds and ends: The coal cost of $27.10 per megawatt hour was reported in TVA's annual SEC report for the fiscal year that ended in September 2017. The letter from the Arkansas delegation—"antithetical to your distinguished record"—is dated March 7, 2017.

16 – Wind Catcher

The White House meeting is described in a briefing memo prepared for federal officials and was confirmed to me be people in attendance.

Senator Alexander's floor speech is available at https://www.you tube.com/watch?v=qxqfoA585mE (accessed July 2018). The Southern Alliance for Clean Energy released details of its investigation into TVA spending on airplanes in early 2018. Coverage of the report can be found in *Utility Dive*, including an article by Robert Walton on February 13, 2018.

The federal government estimates the annual "operating expenses" of running a variety of power plants in table 8.4 of its Electric Power Annual. The annual release in December 2017, revised May 2018, reports 25.36 mills per kilowatt hour, which is equivalent to dollars per megawatt hour.

The Plains & Eastern 53 percent capacity factor can be found in the administrative law judge's decision in the Texas Public Utility Commission case No. 47461, filed May 18, 2018. The nearest wind farm to Wind Catcher/States Edge is the Goodwell Wind Farm. It is no more than thirty miles to the south. In 2016, it generated 860,932 megawatt hours out of a potential 1.75 million—for a 49.1 percent capacity factor, according to federal data.

I got information about Watts Bar Unit 2 from a letter from the TVA to the Nuclear Regulatory Commission, dated May 12, 2017, "Licensee Event Report 391/2017-002-00," and a second letter from May 22, "Licensee Event Report 391/2017-003-000." The Polsky email is dated March 22, 2017, and was sent to White House officials D. J. Gribbin, Jeremy Katz, and Michael Catanzaro. Terry Roland issued a press release the next day, which is the source of his quote. Details of the subsequent meeting between Bill Johnson, John Wilder, and Skelly were relayed to me by Skelly and confirmed by Johnson.

Details of Invenergy's dealing with AEP come from several sources. The most helpful was testimony in the subsequent Oklahoma Corporation Commission's review of the Wind Catcher Energy Connection application, cause PUD 201700267. Paul Chodak's testimony is particularly instructive. Interviews with Carroll Beaman, Antonio Smyth, and Mark McCullough provided additional details. Mario Hurtado's testimony in December provided the 60 percent leased figure and also the "eight years focused on a very similar goal" quote. Details of the July 26 announcement are from press releases and Jay Godfrey's testimony on September 21. The email where Nick Akins said "the reputation and emerging brand of AEP" is part of the "Attorney General's Ninth Data Request" in the Wind Catcher proceedings. The AEP conference call was held on July 27, 2017.

The Arkansas application is docket 17-038-U before the Arkansas Public Service Commission. The Missouri decision is in the case EA-2016-0358 and was issued August 16, 2017. The tweet from Skelly referenced in this chapter is https://twitter.com/CleanLineEnergy/

status/897944837349036032 (accessed July 2018). I reviewed Jimmy Glotfelty's email that is the source of his "rollercoaster of emotions" quote. The same goes for Skelly's "Merry Christmas!" email to Bill Johnson and Cary Kottler's "holds the door open" email.

The order by Judge D. P. Marshall Jr. was in the *Downwind v. U.S. Department of Energy* lawsuit (case 3:16-cv-00207-JLH). The order was signed and filed on December 21, 2017. Bill Johnson's response to the end of the affair was in an article by Dave Flessner that appeared online on December 31, 2017 on the website of the *Chattanooga Times Free Press*.

17 – The Network

My reconstruction of President Eisenhower's trip to Detroit comes from several newspaper sources: a *New York Times* October 30, 1954, article, "Cold, Rain Mark President's Tour," and the *Detroit Free Press*'s coverage of the visit in its October 30 edition—"President Enjoys His Visit to Detroit" and "Jobs Without War, Ike Promises State."

His "shackles of secondary roads" quote is from a transcript of his October 29 speech in Cadillac Square. A transcript is online and prepared by Gerhard Peters and John T. Woolley in *The American Presidency Project*, http://www.presidency.ucsb.edu/ws/?pid=10117 (accessed July 2018).

As noted in chapter 2, the National Academy of Engineering's top twenty list is from *A Century of Innovation*.

The overlay study paper is "Future Cost-Competitive Electricity Systems and Their Impact on US CO_2 Emissions," by Alexander E. Mac-Donald, Christopher T. M. Clack, Anneliese Alexander, Adam Dunbar, James Wilczak, and Yuanfu Xie. It was published online on January 25, 2016, in *Nature Climate Change* (DOI: 10.1038/NCLIMATE2921).

The "there are no computer models sufficient to handle the enormous amount of data" quote is from a Congressional Research Service report, requested by Lee Metcalf and printed in 1976. It is titled

"National Power Grid System Study—An Overview of Economics, Regulatory, and Engineering Aspects."

The best overview of the Cape Wind saga I found was Wendy Williams and Robert Whitcomb's *Cape Wind: Money, Celebrity, Class, Politics, and the Battle for Our Energy Future on Nantucket Sound,* PublicAffairs, 2007. Jim Gordon's quote "hopefully make it easier for other offshore wind developers" is from a press statement he released.

Index